I0494751

DINAMIC GRAVITY

INDUCTION THEORY

Ovidiu Cupsa

**Dynamic Gravity Induction
is
The Missing Part.**

Dynamic Gravity Induction

The basic idea of **DGIT** *is just to accept a simple assumption:*

Gravity is not limited to a central force, but also has a local dynamic inductive influence (DGI).

The stiffness and the adherence of the Space-Time Continuum are the consequences of DGI.

In DGI's presence all the energy systems (massive or not) have the same behaviour!

In my opinion we choose to believe in the existence of some alleged Dark Matter, distributed exactly where we need it for confirming our equations.

If it really exists, let it be! It will definitely become obvious when the Universe will decide.

I agree: it might seem unexciting to use the same old equations as Newton did, and incommodious to suggest adjustments to Einstein's field equations.

Well, I'm not a well-known physicist, and I'd rather have a plain explanation than a complicated theory.

DYNAMIC GRAVITY INDUCTION THEORY
A UNIFIED UNIVERSAL FORCE MODEL

Tesla contradicted Einstein in 1937, claiming that Gravity is not a central force but a dynamic force, having a variable field.

Unfortunately, the mathematical details of Tesla's Dynamic Gravity Theory do not exist. However, Tesla's assumption was based on the existence of an electromagnetic radiation generated by an alleged substance, which he called *ether*.

Dynamic Gravity Induction Theory (DGIT) has in common with Tesla's Dynamic Gravity Theory only the assumption that the gravitational field is a gravitodynamic field, but Dynamic Gravity Induction (DGI) probably has a different interpretation by Tesla. (We will never know!).

According to DGIT, Gravity and EM are both the unique expression of the interaction between Unified Energy of the energy systems.

The Special Cases: <u>Escape Velocity</u> (p 46), <u>White Waves</u> (p 54), <u>G Grey Hole</u> (p 45), <u>G Grey Body</u> (p 65), <u>Black Hole</u> (p 68), <u>Black Point</u> (p 69) confirm the simple unique inductive behaviour.

Furthermore, DGIT does not confute Einstein's energy, admitting the relativistic energy-momentum.

I prefer to present this Unified Universal Force Model in the classical Gravity perspective, but I assume that it could be interpreted departing from any interaction of our Universe.

DGI is a local inductive stress, in the Space-Time Continuum in the presence of dynamic energy systems; depending on their velocity, energy systems have the allowance to move more or less inductive unconstrained in DGI field (they seem to attract or to reject each other).

Gravitational field could be reinterpreted as being a long distance residual EM field; the attraction of the massive bodies could be interpreted as being the expression of a Unified Universal Force which equally affect all the energy systems, massive or not.

For each two energy systems (1) and (2), in a pure DGIT perspective:

$$\frac{p_1{}^2 c^2}{E_1{}^2} = I_1 \frac{E_2 G}{R c^4}$$

Dynamic Elasticity Ratio – The Missing Part

If $I_1 < 1$- the two systems contract the Space-Time.
If $I_1 > 1$ – the two systems expand the Space-Time.

The Fundamental Premise:

We admit that the relativistic energy-momentum conservation law is overall valid in the Universe.

$$E^2 - p^2c^2 = m_0{}^2c^4.$$

The Fundamental Dilemma:

We assume Dark Matter exists and it is responsible for an additional variation of the momentum, but the global energy transfer between Visible Matter and Dark Matter is null.

So, having a variable total momentum and an invariable total energy, how is it possible to conserve the energy-momentum balance in the Universe?

Potential Answer:

> **Dark Matter doesn't really exist, so we have to review certain forces in the Universe. This is Dynamic Gravity Induction: The Missing Part.**

This paper tries to interpret DGI in both classical and relativistic ways; it could be a challenge for the scientists able to quantify DGIT in a more accurate mathematical form.

Abstract

DGI - The Inductive Stiffness of the Space:
DGI describes the local capacity of the gravitational field generated by massive bodies to inductively influence the motion of the other massive bodies in the field.

$$DGI = \frac{mG}{R} \; [m^2/s^2]$$

1. Dynamic Elasticity Ratio:

The Dynamic Elasticity Ratio expresses the inductive capacity of the massive bodies to unconstrainedly move inside of their mutual gravitational field. *inductive elasticity of the space*

$$I_{12} = \frac{\Gamma_{G12}}{G_{12}} = \frac{v_1{}^2 v_2{}^2}{DGI_{12}} = \varsigma_1 \varsigma_2 v_1{}^2 v_2{}^2$$

According to I in our Universe exist:

I = 0 – Black Holes, Black Points,
0 < I ≤ I_c – EG Grey Holes, G Grey Holes,
I_c < I < I_e – Balancing Grey Waves,
I_e < I < ∞ - Expanding Grey Waves and White Waves.

Where: I_c = 1 and I_e = 2.

2. *The Inductive Elasticity Constant:*

> *The Inductive Elasticity Constant of a gravitational system is the characteristic which expresses the predictive local variation of the elasticity depending on the variation of its radius.*
>
> $$\xi = \frac{\varsigma_{max} - \varsigma_{min}}{R} = \frac{1}{mG} \ [s^2/m^3]$$

E.g. Milky Way has:

$$\xi_{MW} = 0.9785 \times 10^{-32} \cong 1 \times 10^{-33} \ [s^2/m^3]$$

The local Dynamic Elasticity Ratio variation depending on the radius of a galaxy is:

$$I_L = \xi_L \omega^2 R_L{}^3 = \varsigma_L \omega^2 R_L{}^2$$

3. *Dynamic Gravity:*

> *Dynamic Gravity expresses the total inductive and conservative gravitational interactions of a combined gravitational field.*
>
> The Missing Part   gravitodynamic factor
> $$DG_{12} = I_{12} G_{12}{}^2 = \frac{v_1{}^2 v_2{}^2}{DGI_{12}} G_{12}{}^2 = \frac{v_1{}^2 v_2{}^2}{G} G_{12} \ [GK^2 m^2/s^4]$$
> kilogals

4. *The Gravitational Balance:*

If: K_1 *and* K_2 *– centrifugal forces;*

> **Inductive Gravitational Balance Law:**
>
> *A massive system sets in an inductive gravitational balance in that specific position in the field where the Dynamic Gravity and the Centrifugal Forces balance each other.*
>
> $$DG_{12} = \underline{I_{12}}G_{12}{}^2 = K_1 \times K_2$$
>
> *The Missing Part*
>
> **In a Conservative Gravitational Balance:**
>
> $$I_{12} = I_c = 1$$

5. *Energy - Momentum Conservation:*

The general DGIT-relativistic equation of the energy-momentum is respected according to bellow.

$$(E_1{}^2 - p_1{}^2c^2)(E_2{}^2 - p_2{}^2c^2) = m_{01}{}^2c^4m_{02}{}^2c^4$$

6. *Universal Energy Systems Models:*

*1. **Open Electromagnetic Systems – White Waves – Feeder Systems** (the electromagnetic waves):*

$$I = \frac{Rc^2}{MG}$$ *– spreading at speed of light if $I > I_e$, they could be deflected by other energy systems if $I > I_e$.*

2. Open Gravitational Systems - G or EG Grey Holes surrounded by Inductive Grey Waves Fields *(unbalanced gravitational fields – e. g. the observable Universe):*

$I_c < I < \infty$ *– expanding systems.*

3. Closed Gravitational Systems *(in a gravitational balance):*

– Closed Inductive Systems - G Grey Holes (in a rigid balance *– e. g. the galaxies):*
$0 < I \leq I_c$*– inductive contracted, having the maximum radius:* $R = \dfrac{MG}{v^2}$.

- Closed Conservative Systems - G or EG Grey Holes surrounded by Conservative Balanced Grey Waves Fields *(e.g. solar systems, planets with satellites):*
$I = I_c$ *in the conservative balance.*

4. Closed Gravitational - Electromagnetic Systems – EG Grey Holes - Massive Systems *(massive bodies: stars, planets, atoms, particles):*

$I \leq I_c$ *(they are electromagnetic and gravitational balanced),* $R \leq \dfrac{MG}{v^2}$.

5. Hub Systems – Black Holes, Black Points - Singularities: $I = 0$ *- contraction tends to R = 0.*

7. The DGIT-relativistic field equation:

$$G^{\alpha\beta} \times \underbrace{D^{\alpha\beta}}_{} = 8\pi T^{\alpha\beta} \times \underbrace{E^{\alpha\beta}}_{}$$

Dynamic Position Factor   *Elastodynamic Factor*

$$\underline{G}^{\alpha\beta} = 8\pi \underline{T}^{\alpha\beta}$$

The relativism represents a reasonable approximation of DGIT-relativism, for the special case of an isotropic Universe (non-inductive space).

8. Some Universal Assumptions of DGIT:

1. The energy systems quantifiable influence the stiffness of the Space and create its adherence.

2. The visible matter is enough to maintain the rigid shape of the Galaxy.

3. The mass, the radius of the galaxies and the position of the stars relative to the center of the galaxies are quantifiable according to DGIT.

4. The expansion, contraction and balance of Universe depend on DGI.

Chapter I
Fundamental premises

A. The Gravity Theories' Limits:

Gravity is the most present and the most misunderstood force of the Universe.

Newtonian theory treats gravitational field as being an interactive field, which induces in the space-time continuum phenomena depending on the mass of massive bodies and the distance between them.

The Gravity is considered a central conservative force, the Hamiltonian, expressed as the amount of the kinetic and the potential energy, constantly remaining in the field.

But, the gravitational behaviour of the massive bodies in the Universe seems to contradict in some situations this assumption.

Newton's equations can't explain how is it possible to keep the balance between Gravity and Centrifugal Force in our Galaxy, even though Gravity decreases depending on square of the radius and Centrifugal Force increases depending on the radius.

In the gravitational balance of a massive body (1) orbiting in a massive body (2) we have:

$$\frac{K_1}{G_{12}} = \frac{m_1 \omega_1^2 R}{\frac{m_1 m_2}{R^2} G} = \frac{\omega_1^2 R^3}{m_2 G}$$

Our Galaxy rotates as a rigid body, having a constant angular velocity and a variable radius, so it seems to be impossible all over to maintain the balance between Gravity and Centrifugal Force.

We also observe that the relativistic equations have no reasonable explanations for the rigid motion of the Galaxy; the relativism admits that the inertial mass varies depending on the speed, but this variation does not supply the necessary mass.

$$G^{\alpha\beta} = \frac{8\pi G}{c^4} \begin{bmatrix} \rho & 0 & 0 & 0 \\ 0 & p & 0 & 0 \\ 0 & 0 & p & 0 \\ 0 & 0 & 0 & p \end{bmatrix}$$

Studying the equation above, we can easily notice that it could be interpreted as an expression of the space-time evolution directly depending on the evolution of the energy-momentum tensor.

But, astronomically we observed that the matter is still not enough to reasonably explain the present architecture of the Universe.

The missing part was replaced with some huge quantities of an alleged Dark Matter.

But the gap still persists, in my opinion.

If we assume Dark Matter exists, than it should be responsible for an additional variation of the Universal momentum, but the global energy transfer between Visible Matter and Dark Matter is null.

So, how is it possible to conserve the energy-momentum balance in the Universe?

According to the relativistic theory we know:

$$E^2 - p^2 c^2 = m_0{}^2 c^4.$$

But an increasing momentum should also proportionally influence the amount of the Universal energy.

Therefore, I refuse to think that the missing part is outside of our visible Universe.

Further, DGIT will assume that:

> **The Gravity of the massive bodies is a local variable, not only depending on the central Gravity, but also depending on their local inductive potential and on their own motion.**

B. The Unified Universal Force

We introduce our study premises based on the general observation that the energy is the interactive element in our Universe.

The total energy of an energetic system is permanently constant:

$$E = m_0 c^2$$

According to our observations, most of the known forces interact in an inverse proportion with the distance (their mutual interaction between two interactive systems being in an inverse proportion with the square of distance).

So, for each of the existing energy systems in the Universe, we can define an intrinsic total interactive potential.

We assume this potential as being responsible for a Unified Universal Force as bellow:

1. The Unified Universal Force of an energy system is a conservative force which expresses the total interactions of that system with the entire Universe.

$$\Gamma_T = \frac{E}{R} = \frac{m_0 c^2}{R}$$

2. The Unified Universal Force can be defined as being the amount of all the Massive interactions (gravitational, inertial, accelerating forces) and EM interactions (electric, magnetic, nuclear forces) of an energy system.

$$\Gamma_T = \Gamma_G + \Gamma_E$$

3. Massive and EM interactions can be both interpreted as being the expression of the general motion of an energy system.

$$\Gamma_G = \frac{m_0 v^2}{R}$$

$$\Gamma_E = \frac{m_0(c^2 - v^2)}{R}$$

We consider:

- *the massive energy fractal* of a system:

$$T = \frac{\Gamma_G}{\Gamma_T} = \frac{v^2}{c^2}$$

- *the electromagnetic energy fractal* of a system:

$$U = \frac{\Gamma_E}{\Gamma_T} = \frac{c^2 - v^2}{c^2}$$

The absolute invariant expression of the energy-momentum conservation in the Universe:

$$\begin{cases} \Gamma_T = U\Gamma_T + T\Gamma_T \\ \quad U + T = 1 \end{cases}$$

U and T are both variable, depending on the motion in the field, but their amount remains permanently constant.

According to this, we can reconsider the difference between EM forces and the Gravity.

Both EM and Gravity similarly participate in various proportions to the energy transfers and they could interchange energy fractals.

However, we consider Gravity and EM have a unique fundamental interpretation, according to the STEMIONICS Model. *(See REFERENCES.)*

Chapter II
Classical reinterpretation of the motion of massive bodies

A. General Equations:

In the relative motion, the mutual interaction between two energetic systems being in an inverse proportion with the square of distance,

We will consider:

$$\Gamma_{T12} = \Gamma_{T1} \times \Gamma_{T2}$$

$$\Gamma_{T12} = \frac{m_1 m_2 c^4}{R^2}$$

$$\Gamma_{G12} = \Gamma_{G1} \times \Gamma_{G2}$$

$$\Gamma_{G12} = \frac{m_1 {v_1}^2 m_2 {v_2}^2}{R^2}$$

$$\Gamma_{E12} = \Gamma_{E1} \times \Gamma_{E2}$$

$$\Gamma_{E12} = \frac{m_1(c^2 - {v_1}^2) m_2(c^2 - {v_2}^2)}{R^2}$$

B. The General Influence of DGI:

Each massive body generates an own gravitational field.

Generally, we consider that the gravitational field is a conservative one and we accept the Gravity as being a central force, which doesn't depend on the local characteristics of the field and on the dynamics of the bodies.

But we can simply notice that the ratio between the Centrifugal Force and the Gravity is:

$$\frac{K_1}{G_{12}} = \frac{\frac{m_1 v_1^2}{R}}{G \frac{m_1 m_2}{R^2}} = \frac{v_1^2 R}{G m_2}$$

This means it depends on the variable term:

$$\frac{v_1^2 R}{G m_2}$$

So, for a constant angular velocity in our Galaxy, Gravity could not be equal to Centrifugal Force, but to the variable expression below:

$$G_{12} = \frac{G m_2}{v_1^2 R} K_1$$

Naturally, increasing K is not a real force, but just a pseudo-force, which only depends on the inertial motion of a body.

We can assume that gravitational field local varies in a different way of the classic interpretation, meaning that the **gravitons are not simply inert transmitting chargers**.

Dynamic Gravity Induction (DGI) Premise:
The relative motion of the massive bodies in their mutual gravitational field is influenced by local dynamic interactions, depending on their field and on their motion characteristics.

We see in expressions above that only the term $v_1{}^2$ depends on the motion.

The expression:

$$\frac{Gm_2}{R}$$

Only depends on the field's parameters, which describe its **local gravitational interactive potential**.

The **local gravitational interactive potential** of a massive body depends in direct proportion on its mass (**m**) and on the gravitational constant (**G**) and in inverse proportion on the distance in the field (**R**).

Definition of DGI:

DGI describes the local capacity of a gravitational field generated by a massive body to inductively influence the relative motion of the other massive bodies in the field.

$$\mathbf{DGI} = \frac{mG}{R} \; [m^2/s^2]$$

The mutual influence depends on the **combined local gravitational interactive potential** of the bodies which are involved in a mutual gravitational interaction, forming together a *gravitational system*.

Definition of Total DGI:

Total DGI describes the local capacity of the combined gravitational field generated by massive bodies to mutually influence their relative motion.

$$DGI_{12} = \frac{m_1 m_2 G^2}{R^2} \; [m^4/s^4]$$

We saw DGI depends on the gravitational field's local characteristics.

This logically means that the high interactive fields manifest an important DGI and the low interactive ones manifest a weak one.

As a consequence, the motion of the massive bodies could be severely or lightly constrained, depending on DGI.

According to its influence, we will prefer to interpret DGI as being an intrinsic property of the support field: Inductive Stiffness of the Space in the presence of a massive body.

The extreme values of DGI describe special situation:

DGI = 0, when the distance between two massive bodies is ∞, or the mass of at least one of them is 0.

DGI = ∞, when the distance between mass centers of two massive bodies is null or at least one of them has an infinite mass.

As a result, we can consider that the dynamics of massive bodies (mass charges) in a gravitational field is influenced by a kind of local dynamic inductive behaviour.

The influence of the dynamics of a massive body by the gravitational field could be somehow compared with the influence of the dynamics of an electrical charge by the induction of the electromagnetic field. *(See ANEX 1 – 4).*

C. Space Elasticity and Dynamic Elasticity Ratio. Dynamic Gravity:

We saw some fields are more permissive than others to allow the massive bodies to move unconstrainedly inside of them.

We define the **inductive allowance of a field**, as being the **inductive elasticity of the space in the gravitational field**, in indirect proportion with DGI:

$$\varsigma = \frac{1}{DGI} = \frac{R}{mG} \; [s^2/m^2]$$

According to DGIT the Space is inductive and adherent, as well as the Time is relative; in DGIT-relativistic interpretation the Space-Time Continuum creates an anisotropic Universe.

If we consider the motion of a massive body (1) in a gravitational field of another massive body (2), the ratio between the total massive interactions and the gravitational interactions will be:

$$\frac{\Gamma_{G1}}{G_{12}} = \frac{\dfrac{m_1 v_1^2}{R}}{\dfrac{m_1 m_2}{R^2} G} = \frac{v_1^2 R}{m_2 G}$$

So:

$$\frac{\Gamma_{G1}}{G_{12}} = \frac{v_1^2}{DGI_2} = \varsigma_2 v_1^2$$

We can easily see the $\frac{\Gamma_{G1}}{G_{12}}$ ratio is in an inverse proportion with DGI or in a direct proportion with the inductive elasticity of the field.

Depending on this ratio, a massive body is able to easily or hardly evolve in an unconstrained inductive motion in the field.

For a determinate relative motion in the gravitational field, we can define $\frac{\Gamma_{G1}}{G_{12}}$ ratio as being the **dynamic elasticity ratio (I)** of that motion.

Definition of Dynamic Elasticity Ratio:
The Dynamic Elasticity Ratio expresses the inductive capacity of a massive body to unconstrainedly move in a gravitational field.

$$I_1 = \frac{\Gamma_{G1}}{G_{12}} = \frac{v_1^2}{DGI_2} = \varsigma_2 v_1^2$$

The combined **I** for a *two bodies gravitational system* could be defined:

Definition of Total Dynamic Elasticity Ratio:
The Total Dynamic Elasticity Ratio expresses the inductive capacity of massive bodies to unconstrainedly move inside of their mutual gravitational field.

$$I_{12} = \frac{\Gamma_{G12}}{G_{12}} = \frac{v_1^2 v_2^2}{DGI_{12}} = \varsigma_1 \varsigma_2 v_1^2 v_2^2$$

The directly measurable physical expression of Dynamic Elasticity Ratio is:

$$I_1 = \frac{\Gamma_{G1}}{G_{12}} = \frac{\frac{m_1 v_1^2}{R}}{\frac{m_1 m_2}{R^2}G} = \frac{v_1^2 R}{m_2 G}$$

The variation of I_1 depending on the variation of its compounds is:

$$\frac{\delta I_1}{\delta v_1} = \frac{2 v_1 R}{m_2 G}$$

$$\frac{\delta I_1}{\delta R} = \frac{v_1^2}{G}$$

$$\frac{\delta I_1}{\delta v_1 \delta R} = \frac{2 v_1}{m_2 G}$$

$$\Gamma_{G1} = \frac{v_1^2 R}{m_2 G} G_{12}$$

$$I_2 = \frac{\Gamma_{G2}}{G_{12}} = \frac{\frac{m_2 v_2^2}{R}}{\frac{m_1 m_2}{R^2}G} = \frac{v_2^2 R}{m_1 G}$$

The variation of I_2 depending on the variation of its compounds is:

$$\frac{\delta I_2}{\delta v_2} = \frac{2v_2 R}{m_1 G}$$

$$\frac{\delta I_2}{\delta R} = \frac{v_2{}^2}{m_1 G}$$

$$\frac{\delta I_2}{\delta v_2 \delta R} = \frac{2v_2}{m_1 G}$$

$$\boxed{\Gamma_{G2} = \frac{v_2{}^2 R}{m_1 G} G_{12}}$$

If Γ_{G12} represents the combined massive interaction in the gravitational field,

We see that:

$$I_{12} = \frac{\Gamma_{G12}}{G_{12}{}^2}$$

$$I_{12} = \frac{v_1{}^2 v_2{}^2 R^2}{m_1 m_2 G^2}$$

The variation of I_{12} depending on the variation of its compounds is:

$$\frac{\delta I_{12}}{\delta v_1 v_2} = \frac{4v_1 v_2 R^2}{m_1 m_2 G^2}$$

$$\frac{\delta I_{12}}{\delta R} = \frac{2v_1{}^2 v_2{}^2 R}{m_1 m_2 G^2}$$

$$\frac{\delta I_{12}}{\delta v_1 \delta v_2 \delta R} = \frac{8 v_1 v_2 R}{m_1 m_2 G^2}$$

$$\Gamma_{G12} = \Gamma_{G1} \times \Gamma_{G2} = \frac{v_1{}^2 v_2{}^2 R^2}{m_1 m_2 G^2}\, G_{12}{}^2 = \frac{v_1{}^2 v_2{}^2}{G}\, G_{12}$$

In fact, we see that the massive interactions could be interpreted as gravitodynamic interactions.

So, we will redefine them as being equal to the total reaction exerted by the gravitational field toward a massive body: **Dynamic Gravity.**

Dynamic Gravity expresses the total inductive and conservative gravitational interactions of a gravitational field.

$$DG_1 = I_2 G_{12} = \frac{v_2{}^2}{DGI_1}\, G_{12} = \varsigma_1 v_2{}^2 G_{12}$$

$$DG_2 = I_1 G_{12} = \frac{v_1{}^2}{DGI_2}\, G_{12} = \varsigma_2 v_1{}^2 G_{12}$$

$$DG_{12} = I_{12} G_{12}{}^2 = \frac{v_1{}^2 v_2{}^2}{DGI_{12}}\, G_{12}{}^2 = \varsigma_1 \varsigma_2 v_1{}^2 v_2{}^2 G_{12}{}^2$$

gravitodinamic factor

$$DG_{12} = \frac{\overbrace{v_1{}^2 v_2{}^2}}{G}\, G_{12}$$

The massive interactions and Dynamic Gravity are equal.

$$\Gamma_{G12} = DG_{12}$$

gravitodynamic factor

$$\Gamma_{G12} = \overbrace{\frac{v_1^{\,2} v_2^{\,2}}{G}} G_{12} = DG_{12}$$

(See ANEX 1 – 4).

In fact the equality of the two terms represents the expression of the **Third Low of Newton**:

Total interactions exerted by the massive bodies towards the gravitational field is equal to total interactions exerted by the gravitational field toward the massive bodies.

D. DGI in the compound fields:

We assume the total combined DGI for **n** gravitational fields is *(See ANEX 6)*:

$$DGI_{Total} = DGI_1 \times DGI_2 \times \ldots \times DGI_n$$

And the **total dynamic elasticity ratio** is:

$$I_{Total} = I_1 \times I_2 \times \ldots \times I_n$$

E. The rotational motion:

In the rotational motion the speed is expressed depending on the radius and on the angular velocity.

$$v = \omega R$$

$$\Gamma_{G12} = \frac{\omega_1{}^2 R^2 \omega_2{}^2 R^2}{c^4} \, \Gamma_{T12}$$

$$\Gamma_{G12} = \frac{\omega_1{}^2 \omega_2{}^2 R^4}{c^4} \, \Gamma_{T12}$$

We can write:

$$I_1 = \frac{\Gamma_{G1}}{G_{12}} = \frac{\dfrac{m_1 v_1{}^2}{R}}{\dfrac{m_1 m_2}{R^2} G} = \frac{\omega_1{}^2 R^3}{m_2 G}$$

The variation of I_1 depending on the variation of its compounds is:

$$\frac{\delta I_1}{\delta \omega_1} = \frac{2 \omega_1 R^3}{m_2 G}$$

$$\frac{\delta I_1}{\delta R} = \frac{3 \omega_1{}^2 R^2}{m_2 G}$$

$$\frac{\delta I_1}{\delta \omega_1 \delta R} = \frac{6\omega_1 R^2}{m_2 G}$$

$$I_2 = \frac{\Gamma_{G2}}{G_{12}} = \frac{\frac{m_2 v_2^2}{R}}{\frac{m_1 m_2}{R^2} G} = \frac{\omega_2^2 R^3}{m_1 G}$$

The variation of I_2 depending on the variation of its compounds is:

$$\frac{\delta I_2}{\delta \omega_1} = \frac{2\omega_2 R^3}{m_1 G}$$

$$\frac{\delta I_1}{\delta R} = \frac{3\omega_2^2 R^2}{m_1 G}$$

$$\frac{\delta I_1}{\delta \omega_1 \delta R} = \frac{6\omega_2 R^2}{m_1 G}$$

$$I_{12} = I_1 \times I_2 = \frac{\omega_1^2 \omega_2^2 R^6}{m_1 m_2 G^2}$$

The variation of I_{12} depending on the variation of its compounds is:

$$\frac{\delta I_{12}}{\delta \omega_1 \delta \omega_2} = \frac{4\omega_1 \omega_2 R^6}{m_1 m_2 G^2}$$

$$\frac{\delta I_{12}}{\delta R} = \frac{6\omega_1{}^2\omega_2{}^2 R^5}{m_1 m_2 G^2}$$

$$\frac{\delta I_{12}}{\delta \omega_1 \delta \omega_2 \delta R} = \frac{24\omega_1 \omega_2 R^5}{m_1 m_2 G^2}$$

$$\Gamma_{G12} = I_{12} G_{12}{}^2 = G_{12}{}^2$$

$$\Gamma_{G12} = \frac{\omega_1{}^2 \omega_2{}^2 R^6}{m_1 m_2 G^2} G_{12}{}^2 = \frac{\omega_1{}^2 \omega_2{}^2 R^4}{G} G_{12}$$

At the same time we can write:

$$\Gamma_{G1} = \frac{m_1 \omega_1{}^2 R^2}{R} = m_1 \omega_1{}^2 R$$

$$\Gamma_{G1} = K_1$$

Where K_1 is the centrifugal force of the body 1.

$$\Gamma_{G2} = \frac{m_2 \omega_2{}^2 R^2}{R} = m_2 \omega_2{}^2 R$$

$$\Gamma_{G2} = K_2$$

Where K_2 is the centrifugal force of the body 2.

$$\Gamma_{G12} = \Gamma_{G1} \times \Gamma_{G2} = \frac{m_1 v_1{}^2 m_2 v_2{}^2}{R^2}$$

$$\Gamma_{G12} = \frac{m_1 \omega_1{}^2 R^2 m_2 \omega_2{}^2 R^2}{R^2}$$

$$\Gamma_{G12} = m_1 \omega_1{}^2 R \times m_2 \omega_2{}^2 R = K_1 \times K_2$$

In practical situations in a gravitational balance:

$$m_1 \omega_1{}^2 R \equiv m_2 \omega_2{}^2 R$$

When massive body (1) is negligible compared to massive body (2), we are only interested of:

The astronomical measurable value of Dynamic Gravity:
$$DG_2 = I_1 G_{12} = K_1$$

F. The value of the Conservative Gravity between two massive bodies:

In the absolute form the Conservative Gravity can be described depending on the Unified Universal Force:

$$\Gamma_{T12} = T\Gamma_{G12} = \frac{c^4}{v_1{}^2 v_2{}^2} \Gamma_{G12}$$

$$\Gamma_{G12} = I_{12}G_{12}$$

$$\Gamma_{T12} = TI_{12}G_{12}$$

$$\Gamma_{T12} = \frac{c^4}{v_1{}^2 v_2{}^2} \frac{v_1{}^2 v_2{}^2 R^2}{m_1 m_2 G^2} G_{12} = \frac{c^4 R^2}{m_1 m_2 G^2} G_{12}$$

$$\Gamma_{T12} = \frac{c^4}{\underbrace{G}} G_{12}$$

gravitodynamic factor at the speed of light

So, we can write the absolute expression of the Gravity in the Universal energetic field as:

The absolute value of the Conservative Gravity is:

$$G_{12} = \frac{G}{c^4}\, \Gamma_{T12}$$

This equation confirms the hypothesis that in the absolute form G_{12} is a conservative force which preserves the total energy-momentum of an energetic system.

In the absolute form, the Dynamic Gravity depending on the Unified Universal Force remains:

$$DG_{12} = \Gamma_{G12} = \frac{v_1{}^2 v_2{}^2}{G} G_{12} = \frac{G}{c^4} \frac{v_1{}^2 v_2{}^2}{G} \Gamma_{T12}$$

Depending on the Dynamic Gravity and total massive interactions, Conservative Gravity is:

$$G_{12} = \frac{Gm_2}{v_1{}^2 R} \Gamma_{G1} = \frac{1}{I_2} DG_1$$

$$G_{12} = \frac{Gm_1}{v_2{}^2 R} \Gamma_{G2} = \frac{1}{I_1} DG_2$$

$$G_{12}{}^2 = \frac{G^2 m_1 m_2}{v_1{}^2 v_2{}^2 R^2} \Gamma_{G12} = \frac{1}{I_{12}} DG_{12}$$

$$G_{12} = \frac{G}{v_1{}^2 v_2{}^2} \Gamma_{G12} = \frac{G}{v_1{}^2 v_2{}^2} DG_{12} = \frac{Gm_1 m_2}{R^2}$$

(See ANEX 1 – 4).

G. The Gravitational Balance:

We saw that the dynamics of massive bodies in a gravitational field is determined by DGI.

In the gravitational field, the massive bodies evolve towards the position in the field where they are in a gravitational balance:

$$G_{12} = \frac{G}{v_1{}^2 v_2{}^2} DG_{12} = \frac{Gm_1 m_2}{R^2} \quad (1)$$

So:

$$\frac{G}{v_1^2 v_2^2} \frac{m_1 v_1^2}{R} \times \frac{m_2 v_2^2}{R} = \frac{m_1 m_2}{R^2} G$$

And:

$$G_{12}^{\,2} = \frac{1}{I_{12}} DG_{12} \ (2)$$

Departing from the equations (1) and (2), we have the Equations system bellow:

$$\begin{cases} G_{12} = \dfrac{G}{v_1^2 v_2^2} DG_{12} \ (1) \\[3mm] G_{12}^{\,2} = \dfrac{1}{I_{12}} DG_{12} \ (2) \end{cases}$$

Simplifying the system:

$$G_{12} = \frac{1}{I_{12}} \frac{v_1^2 v_2^2}{G}$$

$$G_{12} = \frac{1}{I_{12}} \frac{\omega_1^2 \omega_2^2 R^4}{G}$$

The local variation of G_{12} depending on the variation of its compounds will be:

In linear motion:

$$\frac{\delta G_{12}}{\delta v_1 \delta v_2} = \frac{1}{I_{12}} \frac{4v_1 v_2}{G}$$

In rotational motion:

- For constant radius:

$$\frac{\delta G_{12}}{\delta \omega_1 \delta \omega_2} = \frac{1}{I_{12}} \frac{4\omega_1 \omega_2 R^4}{G}$$

- For constant angular velocity:

$$\frac{\delta G_{12}}{\delta R} = \frac{1}{I_{12}} \frac{4\omega_1{}^2 \omega_2{}^2 R^3}{G}$$

- For the variation of all compounds:

$$\frac{\delta G_{12}}{\delta \omega_1 \delta \omega_2 \delta R} = \frac{1}{I_{12}} \frac{16\omega_1 \omega_2 R^3}{G}$$

But:

$$G_{12} = \frac{1}{I_{12}} \frac{\omega_1{}^2 \omega_2{}^2 R^4}{G} = \frac{1}{I_{12}} \frac{\dfrac{K_1 \times K_2 \times R^2}{m_1 m_2}}{\dfrac{G_{12} R^2}{m_1 m_2}}$$

Reducing:

$$G_{12} = \frac{1}{I_{12}} \frac{\omega_1{}^2 \omega_2{}^2 R^4}{G} = \frac{1}{I_{12}} \frac{K_1 \times K_2}{G_{12}}$$

So:

$$G_{12}{}^2 = \frac{1}{I_{12}} (K_1 \times K_2)$$

Or:

$$DG_{12} = I_{12} G_{12}{}^2 = K_1 \times K_2$$

As a conclusion we can state:

Inductive Gravitational Balance Law:
A massive system sets in an inductive gravitational balance in that specific position in the field where the Dynamic Gravity and the Centrifugal Forces balance each other.

$$\textbf{\textit{DG}}_{12} = \textbf{I}_{12}\textbf{G}_{12}{}^2 = \textbf{K}_1 \times \textbf{K}_2$$

Similar to previous calculation:

A massive system sets in a Conservative Gravitational Balance when Dynamic Elasticity Ratio is conservative:
$$I = I_c = 1$$

$$G_{12} = \frac{\omega_1{}^2 \omega_2{}^2 R^4}{G}$$

Conservative Gravitational Balance:

A massive system sets in a conservative gravitational balance when:

$$DG_{12} = G_{12}{}^2 = K_1 \times K_2$$

$$\frac{\omega_1{}^2 \omega_2{}^2 R^4}{G} G_{12} = K_1 \times K_2$$

We will have:

The correspondence between Dynamic Gravity and Conservative Gravity in the Conservative Balance is:

$$DG_{12} = G_{12}{}^2 = \frac{\omega_1{}^2 \omega_2{}^2 R^4}{G} G_{12}$$

$$G_{12} = \frac{\omega_1{}^2 \omega_2{}^2 R^4}{G}$$

We easily can see that the dynamic gravitational balance has a different interpretation than the classical one. *(See ANEX 1 – 4).*

According to the assumptions above, we will further study the dynamics of celestial bodies in the Universe and we will verify if they move according to DGIT.

Chapter III
The Dynamics of Celestial Bodies

A. The common case of the celestial bodies in rotational motion:

We consider a celestial body (1) having mass **m**, which moves on a gravitational orbit around another body (2) having mass **M**:

In the classical expression:

$$m\omega^2 r = G\frac{Mm}{r^2}$$

According to Dynamic Gravity:

$$DG_{12} = K_1 \times K_2$$

$$DG_1 = K_2$$

$$DG_2 = K_1$$

So:

$$\frac{DG_2}{K_1} = I_1 \frac{G_{12}}{K_1}$$

$$I_1 = \frac{K_1}{G_{12}} = \frac{\frac{mv_1^2}{R}}{\frac{MmG}{R^2}} = \frac{v_1^2 R}{MG}$$

$$K_1 = \frac{v_1^2 R}{MG} G_{12}$$

$$DG_2 = I_1 G_{12} = K_1$$

B. The Gravitational Balance of the celestial bodies:

A celestial body which moves on a gravitational orbit is in a gravitational balance when Dynamic Gravity balances Centrifugal Force.

$$DG_2 = I_1 G_{12} = K_1$$

For any **I** value, the system will evolve towards an inductive or a conservative gravitational balance in the field.

To achieve a conservative gravitational balance it must be fulfilled the condition:

$$I_1 = \frac{K_1}{G_{12}} = \frac{\frac{mv_1^2}{R}}{\frac{MmG}{R^2}} = \frac{v_1^2 R}{MG} = 1 = I_c$$

$$DG_2 = G_{12} = K_1$$

When $I \neq 1$, celestial bodies move on apparent orbits different of their conservative orbits, so the general relative motion expression is:

$$R = I\frac{MG}{v^2} = Ir$$

For the rotational motion we will have:

$$\frac{m\omega^2 R}{m\omega^2 r} = I$$

$$R^3 = I\frac{MG}{\omega^2} = Ir^3$$

Due to the action and the reaction, we have:

$$DG_{12} = I_{12}G_{12}^2 = K_1 \times K_2$$

C. Special Case: The Escape Velocity:

$$v_e = \sqrt{\frac{2MG}{R}}$$

$$v_e^2 = \frac{2MG}{R}$$

$$\frac{v_e^2 R}{2MG} = 1 = I_c$$

$$\boxed{\frac{v_e^2 R}{MG} = I_e = 2 \ \textit{(Escape I)}}$$

$$\boxed{\frac{c^2 R}{MG} = I_e = 2 \ \textit{(Light Escape I)}}$$

Escape Dynamic Elasticity Ratio is double of the Conservative Dynamic Elasticity Ratio.
$$I_e = 2I_c = 2$$

D. The Inductive Elasticity Constant of a gravitational system:

1. **For $I > I_c$; $K = IG > G$** – elastic system.
We have an expansion up to:
- to $R = I\frac{MG}{v^2}$, if $I_c < I < I_e$,
- to infinite, if $I \geq I_e$.

2. **For $I = I_c$, $K = IG = G$** – balanced system.
We have a balance: $R = \frac{MG}{v^2}$.

3. **For $I < I_c$, $K = IG < G$** – rigid system.
We have a contraction up to $R = I\frac{MG}{v^2}$.

We can calculate ς variation of a system:

47

$$\frac{\delta\varsigma}{\delta R} = \frac{1}{MG}$$

In a massive system, for the maximum radius, we will have:

$$\varsigma_{max} = \frac{R}{MG}$$

In the immediate vicinity of the black hole:

$$\varsigma_{min} = \frac{0}{MG} = 0$$

The Inductive Elasticity Constant of a gravitational system is the characteristic which expresses the predictive local variation of the elasticity depending on the variation of its radius.

$$\xi = \frac{\varsigma_{max} - \varsigma_{min}}{R} = \frac{1}{MG} \quad \left[\frac{s^2}{m^3}\right]$$

$(\kappa = \frac{1}{\xi} = MG \quad \left[\frac{m^3}{s^2}\right] - $ *Inductive Stiffness Constant*)

Local Inductive Elasticity variation of an inductive gravitational system depending on its own radius is:

$$\xi_L = \xi R_L = \varsigma_L \quad \left[\frac{s^2}{m^2}\right]$$

$$(\kappa_L = \frac{1}{\xi_L} = DGI \left[\frac{m^2}{s^2}\right] \text{- } \textbf{\textit{Local Inductive Stiffness}})$$

We can express local I depending on ξ or on ς:

$$I_L = \xi_L v^2 R_L = \xi_L \omega^2 R_L{}^3$$

$$I_L = \varsigma_L v^2 = \varsigma_L \omega^2 R_L{}^2$$

We consider the local variation of I is responsible for the spiraled shape of the Galaxy. *(See ANEX 7).*

E. Universal Energy Systems Models:

The dynamics of the energy systems is influenced by the value of **DGI** and **I**, they evolve towards the inductive balance. *(See ANNEX 8.)*

I. Depending on DGI we will classify the Universal Energy System as being:

1. White Systems:

- **White Waves** – having EM DGI, null G, finite electromagnetic energy, they spread having light speed through the Space (electromagnetic waves).

- **White Holes** – null DGI, null G, they perfectly reflect the energy, having apparently null proper energy. (Nullons – *See REFERENCES*).

2. Grey Systems:

- **Grey Waves** – having a weak decreasing DGI, they are gravitational fields situated out of the limit of Grey Holes, constraining massive bodies having $1 < I < 2$ to evolve to conservative balanced orbits, or attracting massive bodies having $I < 1$.

- **Grey Holes** – having an important DGI, they are massive bodies or even mass distribution systems of massive bodies, in the captivity of Black Bodies; to their edges the mass partitions have $I \leq 1$.

3. Black Systems:

- **Black Points** – singularities situated in the center of EG Grey Bodies having theoretically infinite DGI and infinite G, null mass, $I = 0$ inside them; they have the behaviour of a Black Hole having a null horizon.

- **Black Holes** – singularities situated in the center of G Grey Bodies, having infinite DGI, infinite G, $I = 0$ inside of them.

II. Depending on I we have:

1. Closed Systems:
- Conservative Balanced,
- Inductive Balanced.

2. Open Systems:
- Inductive Unbalanced.

III. Depending on I and DGI we have:

1. Open Electromagnetic Systems – White Waves – Feeder Systems (the electromagnetic waves - they spread at the speed of light.). In a Grey Waves Field, their $I = \dfrac{Rc^2}{MG}$ and their local EM-DGI is: $EM_{DGI} = \dfrac{Rc^2}{\mu G}$. *(See pages 54 – 55)*.

2. Open Gravitational Systems – G or EG Grey Holes surrounded by Inductive Grey Waves Fields (gravitational fields): $I_c < I < \infty$. Expanding systems toward the conservative balance ($I_c < I < I_e$), or to infinite ($I \geq I_e$).

E.g. the observable Universe.

Out of the limit of Grey Holes (which are gravitational conservative balanced at their edge), the elasticity of the space allows the massive bodies having I > 1 to expand till the conservative balance.

The gravitational interactions are present (Grey Waves), but because of the high I, the massive bodies in the gravitational field of the Grey Holes "escape" till to the geodesic where they set in a conservative balance.

3. Closed Gravitational Systems (balanced systems), which could be:

- Closed Inductive Systems - G Grey Holes (in a rigid gravitational balance): $0 < I \leq I_c$; they are contracted up to $R = I\frac{MG}{v^2}$.

E.g. the galaxies, which have the maximum radius: $R = I\frac{MG}{v^2}$.

These systems are in the captivity of a central mass having an infinite gravity (a black hole); their mass density is distributed relatively uniform in the entire system, in a large number of mass parts which gravitate around the center.

They create the huge counterweight to the central mass, otherwise the matter could be swallowed by this; because these strong gravitational interactions their behaviour is too rigid to be able to evolve towards a conservative balance.

- Closed Conservative Systems - G or EG Grey Holes surrounded by Conservative Balanced Grey Waves Fields: I = 1 in the conservative gravitational balance.

E.g. solar systems, planets with satellites, which generally have an uneven distribution, being formed by a central mass (a Grey Hole) surrounded of some rare partitions which conservatively balanced gravitate around this.

Because of their relatively small radius, and their relatively weak DGI out of the limits of the Grey Hole, they had enough time since their birth to conservatively balance themselves.

4. Closed Gravitational - Electromagnetic Systems – EG Grey Holes - Massive Systems (massive bodies: stars, planets, possible atomic and subatomic particle): $I < I_c$.

Their expansion is additionally impeded by electromagnetic connections – small distance interactions (electrical, nuclear and chemical) $R < \frac{MG}{v^2}$.

At their edges I < 1, we can affirm they are electromagnetically and gravitationally balanced.

5. Hub Systems:

– Small Hub Systems - Black Points - Centers of the EG Grey Holes: G is finite, **DGI =** ∞, **I = 0** – their contraction tends to R = 0.

- Hub Systems - Black Holes - Singularities: DGI = ∞, G = ∞, I = 0 – their contraction tends to R = 0.

F. Special Case: White Waves:

White Waves follow the Space geodesics, at the speed of light, being deflected by the massive bodies when these are able to inductively constrain them:

$$I = \frac{Rc^2}{MG} = I_e = 2 \Rightarrow R = \frac{2MG}{c^2} \text{ (a Black Hole)}$$

We assume that EM radiations are inductively influenced exactly as massive bodies.

The EM-DGI is that characteristic of White Waves which makes it possible to express in the same way EM Induction and DGI.

If we consider μ_1 and μ_2 – the local density of the energy of two different White Waves, according to DGIT, we assume that:

> **The local EM-DGI of a White Wave is:**
> $$EM_{DGI} = \frac{Rc^2}{\mu G}$$

> **A White Wave is able to deflect the other's White Wave geodesic if it has a local energy density able to inductively constrain the other:**
> $$I_1 = \frac{Rc^4}{\mu_2 G} \leq I_e$$

G. The expression of Dynamic Gravity as the amount of Conservative Gravity and DGI:

For the motion of a celestial body (1) having mass **m** in the gravitational field of a celestial body (2) having mass **M**, if **m** is negligible compared to **M**, the body (2) motion is negligible too.

We are only interested in body (1) motion.

In a general linear motion of a massive body (1) in a gravitational field (2):

$$DG_2 = I_1 G_{12} = \frac{v_1^2}{DGI_2} G_{12} = \left(\frac{v_1^2}{DGI_2} - 1\right) G_{12} + G_{12}$$

In a motion having a specific angle α with the field lines:

$$DG_2 = I_1 G_{12} = \frac{v_1^2}{DGI_2} G_{12} = \left(\frac{v_1^2}{DGI_2} - \frac{1}{\cos^2 \alpha} \right) \cos^2 \alpha \, G_{12} + G_{12}$$

E.g. we see that in a free fall:

$$\text{If } \alpha \rightarrow 0, \cos \alpha \rightarrow 0.$$

$$DG_2 \rightarrow G_{12}$$

E.g. for a body (1) in another rotational motion around a body (2):

$$DG_2 = I_1 G_{12} = \frac{\omega_1^2 R^2}{DGI_2} G_{12} = \left(\frac{\omega_1^2 R^2}{DGI_2} - 1 \right) G_{12} + G_{12}$$

We see that if we have a conservative balance:

$$I_1 = 1$$

$$DG_2 = G_{12} = K_1$$

In this case the behaviour of the bodies is exactly described by the classical theory.

> **Right-Hand Rule** - **Direction of DG_2:**
> **1. $I < 1$: thumb toward the center of the field (2).**
> **2. $I > 1$: thumb opposed to center of the field (2).**

H. The Rigid Balance Model of a Galaxy:

We generally accept that black holes oppose disintegration of the galaxies and they release Hawkins radiations, permanently recycling the complex architecture of the Universe.

The theoretic gravity of a black hole is infinite.

$$G_{12} = \frac{m_1 m_2}{R^2} G \rightarrow \infty$$

For a massive body which is in a rotational motion in the immediate vicinity of its horizon, the gravitational balance can be achieved either if the speed of the massive body tends to infinite, either if the mass of the body is null (this means it isn't a massive body anymore).

We should have:

$$G_{12} = K_1$$

So:

$$K_1 = \frac{m_1 \omega_1^2}{R} \rightarrow \infty$$

According to conservation of the momentum, we need a supplementary acceleration to increase the speed of a body which is in an inertial motion.

The massive bodies in the gravitational balance are in an inertial motion, but a supplementary acceleration doesn't exist.

So, the initial momentum has to be preserved.

But, if the Hamiltonian constantly remains and the massive interactions are:

$$\Gamma_{G1} = \frac{m_1 v_1{}^2}{R} = m_1 \omega_1{}^2 R$$

Γ_{G1} is invariable in the gravitational balance.

In fact, we can easily see that:

$$DG_2 = I_1 G_{12} = \frac{v_1{}^2}{DGI_2} G_{12}$$

Could be written as:

$$DG_2 = \underbrace{\left(\frac{R}{m_{2G}} \times \frac{m_{2G}}{R} \right)}_{(equal\ to\ 1)} \frac{m_1 v_1{}^2}{R}$$

Where:

$\dfrac{R}{m_{2G}} = \dfrac{1}{DGI_2}$ – the inductive elasticity of space,

$\frac{m_2 G}{R}$ – gravitational interactive potential.

We can presume the inductive elasticity locally adjusts the central gravity in the gravitational field.

In fact this seems to be the cause of the conservation of the momentum in the field and consequently motion only depend on bodies' inertia.

Therefore, the gravitational field is not responsible for a supplementary momentum, but only influences the direction of the velocity vector.

The inductive elasticity adjusts Gravity, its apparent effect being similar to a kind of pseudo anti-gravity; otherwise the inductive balanced rotational bodies would fall in the black hole, under the influence of its huge gravity.

Inside the black hole, DGI is infinite and the elasticity is null. It has a perfect rigid static center.

In the immediate vicinity of the center, DGI tends to infinite, consequently space is very rigid, drastically reducing the effect of the central Gravity.

This rigidity hardly permits the compression and elongation of the space in immediate vicinity of the black hole.

So, in the vicinity of a huge mass (black-hole) we have a very rigid balance; the massive bodies have to set themselves on some orbits strongly constrained by the influence of the DGI.

As we move away from the center, DGI's influence is reduced, its constraint decreases and the elasticity of the space increases.

We could naturally accept also the relativistic interpretation, that so called Gravity is just a space-time curvature, but we assume the presence of a dynamic energy system in the field inductively modifies the field's isotropy.

DGIT-relativistic we prefer to interpret that the Space has a local inductive elasticity and the space-time deformation not only depends of the energy and momentum, but also of this elasticity.

The phenomenon is similar to the rotational motion of a load hanged on an elastic string. The elasticity of the string doesn't modify the angular velocity, but it causes the string's elongation, which determines the size of the radius.

Consequently the inductive balanced stars have to set on apparent lower orbits than their classical ones. *(See ANEX 1 – 4).*

Chapter IV
DGIT energy-momentum

A massive system, which moves on a gravitational orbit around another body, is in gravitational balance if:

$$\Gamma_{G1} = I_1 G_{12} = K_1$$

$$I_1 = \frac{\Gamma_{G1}}{G_{12}} = \frac{\dfrac{m_1 v_1^2}{R}}{\dfrac{m_1 m_2}{R^2} G} = \frac{v_1^2 R}{m_2 G}$$

$$\frac{\Gamma_{G1}}{G_{12}} = \frac{\dfrac{m_1 v_1^2}{R}}{\dfrac{m_1 m_2}{R^2} G} = \frac{m_1 v_1^2 R}{m_1 m_2 G} = \frac{\dfrac{m_1^2 v_1^2 R}{m_1}}{\dfrac{m_1^2 m_2 G}{m_1}}$$

$$\frac{\Gamma_{G1}}{G_{12}} = \frac{m_1^2 v_1^2 R}{m_1^2 m_2 G}$$

Departing from the energy and the momentum general expressions:

$$p = mv$$

$$E = m_0 c^2$$

$$I_1 = \frac{\Gamma_{G1}}{G_{12}} = \frac{p_1{}^2 R}{\frac{E_1{}^2}{c^4}\frac{E_2}{c^2}G} = \frac{p_1{}^2 c^2}{E_1{}^2}\frac{Rc^4}{E_2 G}$$

DGIT dynamic-inductive expressions of energy-momentum:

$$\frac{p_1{}^2 c^2}{E_1{}^2} = I_1 \frac{E_2 G}{Rc^4}$$

$$\frac{p_2{}^2 c^2}{E_2{}^2} = I_2 \frac{E_1 G}{Rc^4}$$

$$E_1{}^2 - p_1{}^2 c^2 = E_1{}^2\left(1 - I_1 \frac{E_2 G}{Rc^4}\right)$$

$$E_2{}^2 - p_2{}^2 c^2 = E_2{}^2\left(1 - I_2 \frac{E_1 G}{Rc^4}\right)$$

The energy-momentum of a compound energy system depends on Dynamic Elasticity Ratio of its components and their relative position in the Space:

$$\frac{p_1{}^2 c^2}{E_1{}^2} + \frac{p_2{}^2 c^2}{E_2{}^2} = I_1 \frac{E_2 G}{Rc^4} + I_2 \frac{E_1 G}{Rc^4}$$

Writing R as: $R = \frac{ImG}{v^2}$,

DGIT dynamic expressions of energy-momentum:

$$\frac{p_1{}^2c^2}{E_1{}^2} = \frac{E_2}{E_1}\frac{v_1{}^2}{c^2}$$

$$\frac{p_2{}^2c^2}{E_2{}^2} = \frac{E_1}{E_2}\frac{v_2{}^2}{c^2}$$

$$E_1{}^2 - p_1{}^2c^2 = E_1{}^2\left(1 - \frac{E_2}{E_1}\frac{v_1{}^2}{c^2}\right)$$

$$E_2{}^2 - p_2{}^2c^2 = E_2{}^2\left(1 - \frac{E_1}{E_2}\frac{v_2{}^2}{c^2}\right)$$

Total energy-momentum equation will be:

$$\frac{p_1{}^2c^2}{E_1{}^2}\frac{p_2{}^2c^2}{E_2{}^2} = \frac{E_2}{E_1}\frac{v_1{}^2}{c^2}\frac{E_1}{E_2}\frac{v_2{}^2}{c^2}$$

The energy-momentum of a compound energy system directly depends on the dynamics of its components:

$$\frac{p_1{}^2c^2}{E_1{}^2}\frac{p_2{}^2c^2}{E_2{}^2} = \underbrace{\frac{v_1{}^2v_2{}^2}{c^4}}_{\text{dynamic factor}}$$

DGIT-relativistic energy-momentum:

$$(E_1{}^2 - p_1{}^2c^2)(E_2{}^2 - p_2{}^2c^2) = m_{01}{}^2c^4 m_{02}{}^2c^4$$

Or, relativistic more convenient:

$$(E_1{}^2 - p_1{}^2c^2) + (E_2{}^2 - p_2{}^2c^2) = \frac{E_2}{E_1}m_{01}{}^2c^4 + \frac{E_1}{E_2}m_{02}{}^2c^4$$

We know that:

$$m_0{}^2 = m^2\left(1 - \frac{v^2}{c^2}\right)$$

DGI energy-momentum equations according to the inertial mass:

$$(E_1{}^2 - p_1{}^2c^2)(E_2{}^2 - p_2{}^2c^2) = m_1{}^2c^4\left(1 - \frac{v_1{}^2}{c^2}\right)m_2{}^2c^4\left(1 - \frac{v_1{}^2}{c^2}\right)$$

For $I = I_c = 1$ is also valid:

$$\frac{p_1{}^2c^2}{E_1{}^2} = \frac{E_2 G}{R c^4}$$

$$\frac{p_1{}^2c^2}{E_1{}^2} = \frac{v_2{}^2}{c^2}$$

$$\frac{p_2{}^2c^2}{E_2{}^2} = \frac{v_1{}^2}{c^2}$$

Chapter V
The Gravitational Mass and the Inertial Mass

A. Special Case: G Grey Body:

For an inductive balanced massive body being in relative rotation around itself (a galaxy), we have:

$$E_1 = E_2$$

So:

$$(E^2 - p^2c^2)(E^2 - p^2c^2) = m^2c^4(1 - \frac{v^2}{c^2})m^2c^4(1 - \frac{v^2}{c^2})$$

Classical relativistic relation is valid:

$$E^2 - p^2c^2 = m_0^2c^4 = m^2c^4(1 - \frac{v^2}{c^2})$$

For a closed gravitational system which gravitates around its own center, we can calculate the gravitational mass $(m_G = m)$ depending on inertial mass at the maximum radius:

$$I_1 = I_2 = I_c = \frac{\omega^2 R^3}{m_G G} = 1$$

Equivalent to:

$$m_G = \frac{G}{\omega^2 R^3}$$

But, for a closed gravitational system which is in a gravitational balance, the inertial mass varies depending on the radius.

The mean radius has the value according to equation below:

$$R_{med} = \int_0^R R = \frac{1}{2} R$$

We know that gravitational mass and inertial mass of a system are equivalent.

But, if we calculate the inductive inertial mass (apparent inertial mass) depending on the inductive radius, we obtain:

$$\frac{\omega^2 R^3}{m_G G} = \frac{\omega^2 R_{med}^3}{m_{Ga} G} = \frac{\omega^2 R^3}{8 m_{Ga} G}$$

$$\boxed{\frac{m_{Ga}}{m_G} = \frac{1}{8}; \; m_{Ga} = \frac{G}{8\omega^2 R^3}}$$

So, considering that the inductive inertial mass of a body is influenced by DGI, a galaxy has the real gravitational mass 8 times larger than the inductive rest mass; as a result, we consider:

The Dark Matter is not necessary to explain the gravitational balance of a galaxy.

$$For\ I \neq 1, m_{Ga}\ [kg] = Im_G\ [GK]$$

Where: **1 [kg] = I × 1 [GK] (kilogal).**

Kilogal or **galactic kilogram** is the real gravitational mass unit in our Galaxy. *(See ANEX 5).*

E.g. in our Solar system: 1 [kg] ≅ 0.13 [GK].

Gravitational mass and inertial mass are permanently equal and they vary in an inverse proportion to the radius of the Galaxy.

B. Special case: Black Hole's Horizon:

1. DGIT direct calculus:

$$I = \frac{Rc^2}{mG} \leq I_e = 2 \Rightarrow R_{BH} = \frac{2mG}{c^2}$$

2. DGIT-relativistic calculus:

$$E_1 - p_1{}^2 c^2 = E_1{}^2 \left(1 - \frac{E_2 G}{Rc^4}\right)$$

$$E_1 - p_1{}^2 c^2 = m_{01}{}^2 c^4 = m_1{}^2 c^4 \left(1 - \frac{v_1{}^2}{c^2}\right)$$

Transforming:

$$m_{01}{}^2 = m_1{}^2 c^4 \left(1 - \frac{E_2 G}{Rc^4}\right)$$

And because:

$$R_{med} = \int_0^R R = \frac{1}{2} R$$

We will have, at speed of light:

$$\frac{2m_2 G}{Rc^2} = \frac{v_1{}^2}{c^2} = \frac{c^2}{c^2} = 1$$

For:

$$\frac{2mG}{Rc^2} = 1$$

The Schwartzshield radius is (R_{BH}):

$$R_{BH} = \frac{2mG}{c^2}$$

C. <u>Special Case: Black Point:</u>

Exactly in the center of massive bodies, for an infinite small partition of mass:

When:

$$m = \rho \cdot Vol \rightarrow 0$$

If:

$$\frac{2mG}{R_{BP}c^2} \rightarrow 1$$

Then:

The behaviour of a Black Point is similar to the behaviour of a Black Hole having a small infinite radius.

$$\textcolor{red}{R_{BP} \rightarrow 0}$$

Infinite small partition of the Space could have a small infinite mass, so we assume that:

Black Points creates the adherence of the Space to itself (the Continuum).

NOTE: We could imagine a possible Web Continuum: a continuum field of forces exerted between individual finite particles (quanta). Elementary particles could be in a permanent mutual inductive influence. At the maximum speed of light: $\frac{mG}{R_{BP}c^2} = \frac{EG}{R_{BP}c^4} = 1$. For a theoretical elementary particle (EON – Energy Origin Node, elementary mass/energy ≡ 1 ęon), we could assume an I ratio having finite energy and finite R_{BP}. Due to the infinite number of particles in the field, in order to conserve their momentum, they apparently seem to be in a perpetual chaotic motion.)

Chapter VI
DGIT in accordance to Observable Universe

We will roughly verify the GDI influence in the celestial bodies' motion.

We will consider the mass, the angular velocity and the radius of celestial bodies according to their generally accepted values, based on the astronomical observations.

The low accuracy of calculated values can be improved.

A. The Mass of Milky Way:

Milky Way is an inductive balanced closed gravitational system (G Grey Hole).

$$I = \frac{\omega_{MW}{}^2 R_{MW}{}^3}{M_{MW} G} \leq I_c \Rightarrow M_{MW} = \frac{\omega_{MW}{}^2 R_{MW}{}^3}{G}$$

Calculating the Mass:

$$M_{MW} = \frac{\omega_{MW}{}^2 R_{MW}{}^3}{G}$$

For a full rotation of the galaxy (220 mil. years):

$$\omega_{MW}{}^2 = \left(\frac{6.28}{6.93972 \times 10^{15}}\right)^2 = 0.82 \times 10^{-30}$$

For an estimated radius of 50,000 light years:

$$R_{MW}{}^3 = (5 \times 10^{20})^3 = 125 \times 10^{60} = 12.5 \times 10^{61}$$

$$G = 6.674 \times 10^{-11}$$

$$M_{MW} = \frac{0.82 \times 12.5}{6.674} \times \frac{10^{-30} \times 10^{61}}{10^{-11}} = 1.536 \times 10^{42}$$

$$M_{SUN} = 1.98855 \times 10^{30}$$

$$\boxed{M_{G\,MW} = 0.77 \times 10^{12} M_{SUN} \text{ [kg]}}$$

But the mean radius of the Galaxy is:

$$R_{med\,MW} = \int_0^R R_{MW} = \frac{1}{2} R_{MW}$$

Apparent (visible) mass of the Milky Way is:

$$M_{Ga\,MW} = \frac{\omega^2 R_G{}^3}{G} = \frac{M_{G\,MW}}{8}$$

$$\boxed{M_{Ga\,MW} = 0.096 \times 10^{12} M_{SUN} \text{ (GK)}}$$

According to DGIT, we erroneously assume that overall in the Galaxy the weight is measurable in our local kilograms.

So, the real gravitational mass of the Galaxy is 8 times larger than the apparent (inductive) one and there's enough matter in the Galaxy for maintaining the actual balance; Dark Matter is not necessary.

B. The Radius of Milky Way:

The Milky Way is an inductive balanced closed gravitational system (G Grey Hole).

$$I = \frac{\omega_{MW}{}^2 R_{MW}{}^3}{M_{MW} G} \leq I_c \Rightarrow R_{MW}{}^3 = \frac{M_{MW} G}{\omega_{MW}{}^2}$$

Calculating the radius:

$$R_{MW} = \frac{M_{MW} G}{\omega^2}$$

Or:

$$R_{MW} = \sqrt[3]{\frac{MG}{\omega^2}}$$

The real mass of the Galaxy:

$$M_{MW} = 0.77 \times 10^{12} M_{SUN}$$

We have:

$$R_{MW}^{\;3} = \frac{0.77 \times 10^{12} \times 1.98855 \times 10^{30} \times 6.674 \times 10^{-11}}{0.82 \times 10^{-30}}$$

$$R_{MW}^{\;3} = 1.244 \times 10^{62} \; [m^3]$$

$$R_{MW} = 4.99 \times 10^{20} \; [m]$$

$$\boxed{R_{MW} \cong 50,000 \; [light \; years]}$$

C. The Inductive Elasticity of Milky Way:

For the maximum radius (galaxy radius):

$$\boxed{\varsigma_{max \; MW} = \frac{R}{M_{MW}G} = 0.488 \times 10^{-11} \; [\frac{s^2}{m^3}]}$$

For the minimum radius (immediate vicinity of the center):

$$\varsigma_{max \; MW} = \frac{0}{M_{MW}G} = 0$$

$$\xi_{MW} = \frac{\varsigma_{max \; MW} - \varsigma_{min \; MW}}{R_{MW}^{\;2}} = \frac{1}{M_{MW}G} \; [\frac{s^2}{m^3}]$$

So:

$$\xi_{MW} = 0.9785 \times 10^{-32} \cong 1 \times 10^{-33} \ [s^2/m^3]$$

In a sector of the Galaxy, having a radius R_L the local value of elasticity ξ_{MW} is:

$$\xi_{MWL} = \xi_{MW} R_L = \varsigma_{MW} \ [s^2/m^2]$$

And the local elasticity ratio is:

$$I_{MWL} = \xi_{MW} v^2 R_L = \xi_{MWL} \omega^2 R_L{}^3$$

D. Sun's orbit around Milky Way's center:

The Milky Way is a closed gravitational system inductively balanced (G Grey Hole).

We will have:

$$I_{SUN} = \frac{\omega_{SUN}{}^2 R_{SUN}{}^3}{M_{MW} G} < I_c \Rightarrow R_{SUN}{}^3 = \frac{M_{MW} G}{\omega_{SUN}{}^2}$$

Starting from:

$$I_{SUN} = \frac{\omega_{SUN}{}^2 R_{SUN}{}^3}{M_{MW} G}$$

Theoretically, considering all the mass of the galaxy concentrated in its mass center, we will have:

$$M_{MW} = 0.77 \times 10^{12} M_{SUN}$$

$$I = \frac{2.2 \times 10^5 \times 2.2 \times 10^5 \times 2.7 \times 10^{20}}{0.77 \times 10^{12} \times 1.98855 \times 10^{30} \times 6.674 \times 10^{-11}}$$

$$I_{SUN} \cong 0,13$$

$$R_{SUN}{}^3 = I_{SUN} r_{SUN}{}^3$$

The radius of Sun's inductive orbit (R_{SUN}) compared to the classical one (r_{SUN}) is:

$$R_{SUN} = \sqrt[3]{0.13 r_{SUN}{}^3} \cong 0.5 r_{SUN} \; [\text{m}]$$

We can calculate I_{SUN} also using:

$$I_{SUN} = \xi_{MW}\, \omega_{SUN}{}^2 R_{SUN}{}^3 \cong 0.13$$

And escape velocity could be calculated from:

$$\frac{I_e}{I_{SUN}} \cong \frac{2}{0,13} \cong 15.4$$

$$\frac{M_{SUN}\omega_{SUN}{}^2 R_{ESC}{}^3}{M_{SUN}\omega_{SUN}{}^2 R_{SUn}{}^3} = \frac{R_{ESC}{}^3}{R_{SUN}{}^3} = 15.4$$

$$R_{SUN} = \sqrt[3]{15.4 R_{SUN}{}^3}$$

$$v_{ESC} = 2.48 \, v_{SUN}$$

So, we observe v_{ESC} is similar to the classical observations:

$$v_{ESC} \cong 545.6 \, [km/s]$$

E. Earth's orbit around Sun:

The Solar System is a conservatively balanced closed gravitational system (EG Grey Hole surrounded by Grey Waves Fields).

For the Earth-Moon system, we can calculate the average radius of its orbit around the Sun:

$$I_{SUN} = \frac{\omega_{Earth}{}^2 R_{Earth}{}^3}{M_{SUN} G} = I_c \Rightarrow R_{Earth}{}^3 = \frac{M_{SUN} G}{\omega_{SUN}{}^2}$$

$$R_{Earth} = \sqrt[3]{\frac{M_{SUN}\,G}{\omega^2}} = \sqrt[3]{\frac{1.98855 \times 10^{30} \times 6.674 \times 10^{-11}}{3.96 \times 10^{-14}}} \, [m]$$

So, we will have the mean radius of the Earth-Moon system's orbit around Sun:

$$R_{Earth} \cong 149,650,000 \, [km]$$

F. Milky Way's center rotation around its own mass center:

From the Gravitodynamic Factor:

$$G_{12} = I \frac{v_{MW}^2 v_{SUN}^2}{G} = \frac{M_{MW} m_{SUN} G}{R^2}$$

$$v_{MW}^2 = \frac{(6.674 \times 10^{-11})^2 \times 0.77 \times 10^{11} \times (1.98855 \times 10^{30})^2}{0.13 \times 4.84 \times 10^{10} \times 25 \times 10^{40}}$$

Due to influence of our Sun, the speed of Milky Way's center on its own orbit could be:

$$v_{MW} = 2.31 \, [m/s]$$

A complete rotation of Milky Way lasts 220 million years and in this period, its center could browse a circumference of:

$$C = 9.625 \times 10^{15} \, m$$

In this case, the average radius of the central orbit due to our Sun could be:

$$R_{C\,MW} = 1.53 \times 10^{15} \, m \cong 0.15 \, [ly] \cong 3 \times 10^{-6} \, R$$

However, due to the even distribution of the mass on the Galaxy's disk, we assume that *the geometrical and mass center* is a rigid static one.

G. Sun's motion around Solar System's mass center:

From the Gravitodynamic Factor:

$$G_{12} = \frac{v_{sun}{}^2 v_{Earth}{}^2}{G} = G \frac{M_{SUN} \, m_{Earth}}{R^2}$$

$$v_{Earth} = 29{,}291{,}000 \text{ m/s}$$

$$m_{Earth} = 5.9742 \times 10^{24} \, kg$$

$$M_{SUN} = 1.98855 \times 10^{30} \, kg$$

$$R \cong 149{,}650{,}000{,}000 \text{ m}$$

$$v_{c\,sun}{}^2 = \frac{6.674 \times 10^{-11} \times 6.674 \times 10^{-11} \times 1.98855 \times 10^{30} \times 5.9742 \times 10^{24}}{2.24 \times 10^{22} \times 8.58 \times 10^{10}}$$

$$\boxed{v_{c\,sun} = 52.5 \text{ [m/s]}}$$

During a complete rotation of the Sun (25.38 days) his center browses a distance of about 11.5% of its diameter due to Earth's revolution.

Due to the uneven mass distribution in the solar system, Sun has an uneven motion, so we can't certainly calculate an orbit, only we can estimate its observed position around its mass center.

H. Earth's Inductive Elasticity:

Earth is an EG Grey Hole; if it could be a G Grey Hole, then its radius would be:

$$R_{earth} = \frac{2M_{Earth}\,G}{v_{surface\,med}^2} \cong 2 \times 10^{10}.$$

But real radius is: $R = 6.371 \times 10^6$.

The additional EM complex compound forces reduce the radius about 3140 times more than Gravity alone, which could reconfirm the Gravity is a residual force of the EM forces.

$$\varsigma_{max} = \frac{R_{Earth}}{M_{Earth}\,G} \times \frac{1}{3140} = 0.50 \times 10^{-11}\ [\frac{s^2}{m^3}]$$

Which verifies the exclusively Inductive Elasticity of the Earth is about equal to the Galaxy's:

$$\xi_{earth} = \frac{1}{M_{Earth}\,G} \times \frac{1}{3140} = 0.796 \times 10^{-18}\ [\frac{s^2}{m^3}]$$

Earth's Total Stiffness is more than 3000 times higher than the Galaxy's Stiffness:

$$\frac{DGI_{Earth} + EM\,DGI_{Earth}}{DGI_{MW}} = 3{,}140\ times$$

Chapter VII
DGIT- relativistic interpretation

Further we will infer the proportionality between DGIT and relativistic theory, without claiming a complete approach.

The accuracy of proposed equations could be improved by the scientists who exactly understand the metric of relativistic space-time field, using a more elaborated mathematical formalism.

Once two energy systems interact each other, both introduce in their interaction the influence of their own motion, which is previous and independent of their interaction.

The initial momentum of a massive body is preserved and transferred in the new gravitational field of the dual system.

> **The influence of the proper momentum of an individual moving massive body can be relativistic interpreted as: a dynamic deformation of the space-time – "a turbulence in the laminar space-time field".**

We will depart from the most simple form of the field equation:

$$G^{\alpha\beta} = \frac{8\pi G}{c^4} T^{\alpha\beta}$$

And we will assume that, according to the Dynamic Gravity, the central Gravity is locally adjusted by DGI as below:

$$DG_1 = I_2 G_{12} = \frac{v_2{}^2}{DGI_1} G_{12} = \varsigma_1 v_2{}^2 G_{12}$$

$$DG_2 = I_1 G_{12} = \frac{v_1{}^2}{DGI_2} G_{12} = \varsigma_2 v_1{}^2 G_{12}$$

$$DG_{12} = I_{12} G_{12}{}^2 = \frac{v_1{}^2 v_2{}^2}{DGI_{12}} G_{12}{}^2 = \varsigma_1 \varsigma_2 v_1{}^2 v_2{}^2 G_{12}{}^2$$

$$DG_{12} = \frac{v_1{}^2 v_2{}^2}{G} G_{12}$$

We will prefer the equations:

$$DG_1 = \varsigma_1 v_2{}^2 G_{12}$$

$$DG_2 = \varsigma_2 v_1{}^2 G_{12}$$

$$DG_{12} = \varsigma_1 \varsigma_2 v_1{}^2 v_2{}^2 G_{12}{}^2$$

These equations express DG in accordance to the **inductive elasticity of the space.**

If we consider the relativistic space-time deformation in a compound gravitational field of two massive bodies (1) and (2),

We will have:

$T^{\alpha\beta}$ - stress–energy tensor

Considering an isotropic Universe, we can write the tensor $T^{\alpha\beta}$ for a perfect fluid:

$$T^{\alpha\beta} = \begin{bmatrix} \rho & 0 & 0 & 0 \\ 0 & p & 0 & 0 \\ 0 & 0 & p & 0 \\ 0 & 0 & 0 & p \end{bmatrix}$$

This tensor is the cause of the space-time curvature in the general relativity, replacing the mass density, which is the source of such a gravitational attraction in Newtonian Gravity.

$T^{\alpha\beta}$ depends on the values of matter, radiation, and non-gravitational force fields, and for the compound field we will note them for the moment as $T_{12}{}^{\alpha\beta}$.

In fact, in $T_{12}{}^{\alpha\beta}$ each value of ρ represents the local density in the compound isotropic field.

And each value of p represents the value of the local classic pressure inside of a compound isotropic gravitational field; and we consider it as being **the pressure which expresses the laminar flow of the entire matter 4-current.**

But, according to DGIT, the total interactions (DG) depend also on inductive elasticity of the space and on the dynamics of the bodies.

So, we reconsider these characteristics as being independent variables in the field, which are able to adjust the space-time curvature, as below:

- inductive elasticity, could be considerate an intrinsic local characteristic of the space,

- motion (velocity, acceleration) of the other massive bodies, which are present in the field, can be considered as being an independent local dynamic stress (a turbulence in the laminar matter 4-current).

In an arbitrary frame of reference, considering the Jacobian matrix of an anisotropic (having variable elasticity) space, we can write:

The gradient of the **inductive elasticity of the four-space** in the gravitational field (1) can be written as below:

$$\nabla \varsigma_1 = \begin{bmatrix} \delta \varsigma_1^0 \\ \delta \varsigma_1^x \\ \delta \varsigma_1^y \\ \delta \varsigma_1^z \end{bmatrix}$$

And, if there exists a moving body (2) in the field, the dynamics of this body also locally influences space-time curvature.

Further considering the 4-velocity as being the rate of change of 4-position, with respect to the proper time along the world line of an massive body,

The gradient of 4-velocity of body (2), will be:

$$\nabla U = \begin{bmatrix} \delta v_2^0(\tau) \\ \delta v_2^x(\tau) \\ \delta v_2^y(\tau) \\ \delta v_2^z(\tau) \end{bmatrix}$$

We could rewrite the field equation (introducing for the moment, in the left part, an unknown term $x_1{}^{\alpha\beta}$) as:

$$G^{\alpha\beta} \times x^{\alpha\beta} = \frac{8\pi G}{c^4} T_{12}{}^{\alpha\beta} \times \varsigma_1 u_2{}^2$$

Introducing the ***Elastodynamic Factor*** $(E_2{}^{\alpha\beta})$, depending on the inductive elasticity of the space, and on the velocity of the moving massive body (2) in the gravitational field (1),

We can write the field equation as:

$$G_1{}^{\alpha\beta} \times x_1{}^{\alpha\beta} = \frac{8\pi G}{c^4} T_{12}{}^{\alpha\beta} \times E_2{}^{\alpha\beta}$$

Analyzing the equation in a matrix form, for an anisotropic Universe, we can reconsider the classical form (which is proper for a perfect fluid).

For a general relative motion we will have:

$$G_1{}^{\alpha\beta} \times x_1{}^{\alpha\beta} = \frac{8\pi G}{c^4} \begin{bmatrix} \rho & 0 & 0 & 0 \\ 0 & p & 0 & 0 \\ 0 & 0 & p & 0 \\ 0 & 0 & 0 & p \end{bmatrix} \times \varsigma_1 u_2{}^2$$

We easily see the field equations can be classically expressed if:

$$\varsigma_1 v_2{}^2 = 1$$

But, for rewriting $E_2{}^{\alpha\beta}$ in a more convenient form, we will study the classical equations.

We observe that:

$$G^{\alpha\beta} \sim k\rho$$

Reconsidering ρ in a tensor form:

$$G^{\alpha\beta} \sim kT^{\alpha\beta}$$

Or:

$$G^{\alpha\beta} = \frac{8\pi G}{c^4} T^{\alpha\beta}$$

But, the density means:

$$\rho = \frac{m}{Vol}$$

If we write 4-dimensional mass:

$$m = m\left(\frac{R_0}{\tau} + \frac{R_x}{\tau} + \frac{R_y}{\tau} + \frac{R_z}{\tau} \right)$$

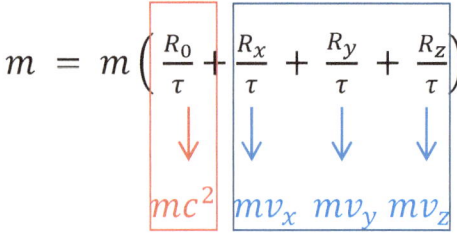

$$mc^2 \quad mv_x \quad mv_y \quad mv_z$$

We can write the expression of ρ:

$$\frac{mc^2}{Vol} = \frac{m}{Vol}c^2 = \rho c^2$$

And of the pressure in the field:

$$\frac{mc^2}{Vol} = \frac{Wk}{Vol} = \frac{\Gamma_T R}{R^3} = \frac{\Gamma_T}{A} = p$$

The 4-velocity of the massive body (2) is a standard 4-vector:

$$U = \gamma(c, u_x, u_y u_z)$$

Analysing the 4-velocity components of the massive body (2):

$$u_2 = \gamma \begin{bmatrix} u_2^0 \\ u_2^x \\ u_2^y \\ u_2^z \end{bmatrix}$$

We will have:

$$u_2^0 = \gamma c$$

And writing the pressure in the field for each axis components, we will have:

$$\frac{mu_2^x}{Vol} = \frac{Wk}{Vol} = \frac{\Gamma_{Gx}R}{R^3} = \frac{\Gamma_{Gx}}{A} = p_{2x}$$

$$\frac{mu_2^y}{Vol} = \frac{Wk}{Vol} = \frac{\Gamma_{Gy}R}{R^3} = \frac{\Gamma_{Gy}}{A} = p_{2y}$$

$$\frac{mu_2^z}{Vol} = \frac{Wk}{Vol} = \frac{\Gamma_{Gz}R}{R^3} = \frac{\Gamma_{Gz}}{A} = p_{2z}$$

For each mass unit, we will have:

$$u_2^x \sim p_{2x}$$

$$u_2^y \sim p_{2y}$$

$$u_2^z \sim p_{2z}$$

Considering ς a property of the space, and the Universe an anisotropic space,

A more accurate form of the relation between velocity and pressure could be calculated departing from Bernoulli's equations.

Considering the components of p_2 **as being the local new pressure in the mass 4-current, determined by the supplementary stress (a turbulence) caused by the motion of massive body (2) in the anisotropic space,**

And p as being the classical proper pressure of the isotropic field,

We could write, in a general 3D form, the local pressure variation generated by the moving body (2), as in Bernoulli equation:

$$p - p_2 = \frac{\rho}{2}\, v_2{}^2$$

But, in relativistic terms, the general expression of the relativistic momentum is:

$$P = \left(\frac{E}{c}, p_x, p_y, p_z\right) = mu$$

Departing from a general 4×4 translation matrix of the Lorenz group, which could be the correspondent of the 4-vector of the square of 4-velocity:

$$U = \begin{bmatrix} u_{11}^2 & u_{12}^2 & u_{13}^2 & u_{14}^2 \\ u_{21}^2 & u_{22}^2 & u_{23}^2 & u_{24}^2 \\ u_{31}^2 & u_{32}^2 & u_{33}^2 & u_{34}^2 \\ u_{41}^2 & u_{42}^2 & u_{43}^2 & u_{44}^2 \end{bmatrix}$$

We can write, in the most simple 4 × 4 matrix form, the corresponding square 4-velocity for a determinate motion, in an arbitrary frame of reference:

$$U_2^{\,2} = 2\gamma^2 \begin{bmatrix} c^2 & 0 & 0 & 0 \\ 0 & \dfrac{p-p_2}{\rho} & 0 & 0 \\ 0 & 0 & \dfrac{p-p_2}{\rho} & 0 \\ 0 & 0 & 0 & \dfrac{p-p_2}{\rho} \end{bmatrix}$$

According to new local pressure in the field,

The elastodynamic factor of the motion of massive body (2) in the field (1) will be:

$$E_2^{\,\alpha\beta} = 2 \begin{bmatrix} c^2 & 0 & 0 & 0 \\ 0 & \dfrac{p-p_2}{\rho} & 0 & 0 \\ 0 & 0 & \dfrac{p-p_2}{\rho} & 0 \\ 0 & 0 & 0 & \dfrac{p-p_2}{\rho} \end{bmatrix} \begin{bmatrix} \varsigma_1^0 \\ \varsigma_1^x \\ \varsigma_1^y \\ \varsigma_1^z \end{bmatrix}$$

In fact we see the equivalence with algebraically 3D form:

$$E_2{}^{\alpha\beta} \equiv \left(c^2, u_2{}^2\right)\left(c^2, \frac{I_2}{u_2{}^2}\right) \equiv \frac{p_2{}^2 c^2}{E_2{}^2} \frac{Rc^4}{E_1 G} = I_2$$

$E_1{}^{\alpha\beta}$ doesn't represent itself a kind of stress-energy expression, able to influence alone the space-time curvature; it represents an elastodynamic adjusting factor, which could not be defined in the absence of the stress-energy tensor.

However, we consider that aplying Elastodynamic factor, we could more accurately describe the space-time curvature.

According to $E_2{}^{\alpha\beta}$,

For the general motion, we could rewrite the field equation (introducing for the moment, in the left part, an unknown term $x_1{}^{\alpha\beta}$) as:

$$G_1{}^{\alpha\beta} \times x_1{}^{\alpha\beta} = \frac{8\pi G}{c^4}\begin{bmatrix} p & 0 & 0 & 0 \\ 0 & p & 0 & 0 \\ 0 & 0 & p & 0 \\ 0 & 0 & 0 & p \end{bmatrix} \times 2\gamma^2 \begin{bmatrix} c^2 & 0 & 0 & 0 \\ 0 & p-p_2 & 0 & 0 \\ 0 & p & p-p_2 & 0 \\ 0 & 0 & p & p-p_2 \\ 0 & 0 & 0 & p \end{bmatrix}\begin{bmatrix} \varsigma_1^0 \\ \varsigma_1^x \\ \varsigma_1^y \\ \varsigma_1^z \\ \varsigma_1^z \end{bmatrix}$$

But, reanalizing Einstein field (ignoring Λ, which is an artificial expression, considered error by Einstein himself), we see:

$$G^{\alpha\beta} = R^{\alpha\beta} - \frac{1}{2} g^{\alpha\beta} R$$

And:

$$G^{\alpha\beta} = \frac{8\pi G}{c^4} T^{\alpha\beta}.$$

But $T^{\alpha\beta}$ energy-stress tensor is divergenceless, and the product from the right part of our new equation admites a divergence.

So, we have to modify the left term of the equation.

But we know that in classic 3D expression:

$$\varsigma_1 = \frac{I_2}{v_2{}^2}$$

Also we have:

$$I_2 = \varsigma_1 v_2{}^2$$

I is a dimensionless term, which expresses the dynamic elasticity ratio of the motion of a massive body, in an anisotropic field.

We can write, in the most simply 4×4 matrix form, the square 4-velocity of massive body (2):

$$u_2 = \gamma^2 \begin{bmatrix} c^2 & 0 & 0 & 0 \\ 0 & u_2^{x2} & 0 & 0 \\ 0 & 0 & u_2^{y2} & 0 \\ 0 & 0 & 0 & u_2^{z2} \end{bmatrix}$$

And, analyzing ς_1 we simply observe, if we write it as:

$$\varsigma_1 = \xi_1 R$$

It could be written as a scalar, which varies depending on the radius.

We can rewrite it in a simple form as a 4 x 1 matrix:

$$\varsigma_1 = \begin{bmatrix} x_1^0 \\ x_1^x \\ x_1^y \\ x_1^z \end{bmatrix}$$

The Dynamic Position Factor (DP Factor) - D_1 could be:

$$D_1{}^{\alpha\beta} = \begin{bmatrix} c^2 & 0 & 0 & 0 \\ 0 & u_2^{x2} & 0 & 0 \\ 0 & 0 & u_2^{y2} & 0 \\ 0 & 0 & 0 & u_2^{z2} \end{bmatrix} \begin{bmatrix} x_1^0 \\ x_1^x \\ x_1^y \\ x_1^z \end{bmatrix}$$

And, simplifying γ^2 in the field equation and unitizing G and c, it could be rewritten as:

$$G^{\alpha\beta} \times \underbrace{D_1{}^{\alpha\beta}}_{DP\ Factor} = 8\pi T_{12}{}^{\alpha\beta} \times \underbrace{E_2{}^{\alpha\beta}}_{Elastodynamic\ Factor}$$

In the same way, for the massive body (1) which moves in the anisotropic field (2) we can follow the same reasoning.

Analysing the 4-velocity components of the massive body (1):

$$U_1 = \begin{bmatrix} u_1^0 \\ u_1^x \\ u_1^y \\ u_1^z \end{bmatrix}$$

Considering the components of **p_1 as being the local new pressure in the mass 4-current, determined by the supplementary stress (a turbulence) caused by the motion of massive body (1) in the anisotropic space,**

And p as being the classical proper pressure of the gravitational isotropic field,

We could, in a general 3D form, write the local pressure variation generated by the moving body (1), as in Bernoulli equation:

$$p - p_1 = \frac{\rho}{2} v_1(\tau)^2$$

Considering the general expression of the relativistic momentum:

$$P = \left(\frac{E}{c}, p_x, p_y, p_z\right) = mu$$

Departing from a general 4 × 4 translation matrix of the Lorenz group, which could be the correspondent of the 4-vector of the square of 4-velocity:

$$U = \begin{bmatrix} u_{11}^2 & u_{12}^2 & u_{13}^2 & u_{14}^2 \\ u_{21}^2 & u_{22}^2 & u_{23}^2 & u_{24}^2 \\ u_{31}^2 & u_{32}^2 & u_{33}^2 & u_{34}^2 \\ u_{41}^2 & u_{42}^2 & u_{43}^2 & u_{44}^2 \end{bmatrix}$$

We can write, in the most simple 4 × 4 matrix form, the corresponding square 4-velocity for a determinate motion, in an arbitrary frame of reference:

$$U_1{}^2 = 2\gamma^2 \begin{bmatrix} c^2 & 0 & 0 & 0 \\ 0 & \dfrac{p-p_1}{\rho} & 0 & 0 \\ 0 & 0 & \dfrac{p-p_2}{\rho} & 0 \\ 0 & 0 & 0 & \dfrac{p-p_2}{\rho} \end{bmatrix}$$

According to the new local pressure in the field,

The Elastodynamic Factor $(E_2{}^{\alpha\beta})$ of the motion of massive body (1) in the field (2) could be written as:

$$E_1{}^{\alpha\beta} = 2 \begin{bmatrix} c^2 & 0 & 0 & 0 \\ 0 & \dfrac{p-p_1}{\rho} & 0 & 0 \\ 0 & 0 & \dfrac{p-p_2}{\rho} & 0 \\ 0 & 0 & 0 & \dfrac{p-p_2}{\rho} \end{bmatrix} \begin{bmatrix} \varsigma_2^0 \\ \varsigma_2^x \\ \varsigma_2^y \\ \varsigma_2^z \end{bmatrix}$$

According to $E_1{}^{\alpha\beta}$,

The field equation can be written depending on the dynamic pressure in the field and on the elasticity of the space.

For the general motion:

We could rewrite the field equation (introducing for the moment, in the left part, an unknown term $x_2{}^{\alpha\beta}$) as:

$$G_2{}^{\alpha\beta} \times x_2{}^{\alpha\beta} = \frac{8\pi G}{c^4} \begin{bmatrix} \rho & 0 & 0 & 0 \\ 0 & \rho & 0 & 0 \\ 0 & 0 & \rho & 0 \\ 0 & 0 & 0 & \rho \end{bmatrix} \times 2\gamma^2 \begin{bmatrix} c^2 & 0 & 0 & 0 \\ 0 & \dfrac{p-p_1}{\rho} & 0 & 0 \\ 0 & 0 & \dfrac{p-p_1}{\rho} & 0 \\ 0 & 0 & 0 & \dfrac{p-p_1}{\rho} \end{bmatrix} \begin{bmatrix} \varsigma_2^0 \\ \varsigma_2^x \\ \varsigma_2^y \\ \varsigma_2^z \end{bmatrix}$$

But, from the classical 3D expression we know that:

$$\varsigma_2 = \frac{I_1}{v_1{}^2}$$

Also we have:

$$I_1 = \varsigma_1 v_2{}^2$$

We can write, in the most simple 4×4 matrix form, the square 4-velocity of massive body (1):

$$u_1 = \gamma^2 \begin{bmatrix} c^2 & 0 & 0 & 0 \\ 0 & u_1^{x2} & 0 & 0 \\ 0 & 0 & u_1^{y2} & 0 \\ 0 & 0 & 0 & u_1^{z2} \end{bmatrix}$$

And, analyzing ς_2 we easily observe, if we write it as:

$$\varsigma_2 = \xi_2 R$$

It could be written as a scalar, which varies depending on the radius:

$$\varsigma_2 = \begin{bmatrix} x_2^0 \\ x_2^x \\ x_2^y \\ x_2^z \end{bmatrix}$$

The Dynamic Position Factor (DP Factor) - D_2 could be:

$$D_2{}^{\alpha\beta} = \begin{bmatrix} c^2 & 0 & 0 & 0 \\ 0 & u_1^{x\,2} & 0 & 0 \\ 0 & 0 & u_1^{y\,2} & 0 \\ 0 & 0 & 0 & u_1^{z\,2} \end{bmatrix} \begin{bmatrix} x_2^0 \\ x_2^x \\ x_2^y \\ x_2^z \end{bmatrix}$$

And, simplifying γ^2 in the field equation and unitizing G and c, it could be rewritten as:

$$G^{\alpha\beta} \times \underbrace{D_2{}^{\alpha\beta}}_{DP\ Factor} = 8\pi T^{\alpha\beta} \times \underbrace{E_1{}^{\alpha\beta}}_{Elastodynamic\ Factor}$$

In fact, considering the mutual field as being a continuous one, described by intrinsic characteristics, we can assume that:

$p - p_1$ and $p - p_2$ **represent both the same expression of a local dynamic variation of the pressure in the field (Δp), caused by local turbulences in the laminar flow of the matter 4-current.**

ς is a proper characteristic of the combined field – local inductive elasticity of the space.

$$G^{\alpha\beta} \times \underbrace{D^{\alpha\beta}}_{DP\ Factor} = 8\pi T^{\alpha\beta} \times \underbrace{E^{\alpha\beta}}_{Elastodynamic\ Factor}$$

This could be interpreted as being equivalent to:

$$G^{\alpha\beta} \times D^{\alpha\beta} = 8\pi I T^{\alpha\beta}$$

We also suggest using the simple notation:

$$\underline{G}^{\alpha\beta} = 8\pi \underline{T}^{\alpha\beta}$$

According to DGIT, we can predict more accurately the evolution of the Universe.

According to I value:
I > 1- we have an expanding Universe,
I = 1- we have a static Universe,
I < 1 – we have a contracting Universe.

We consider, according to I value, two particles could permanently have the ability to inductively influence each other, even though they are in two different edges of the Universe, if the distance between them is finite.

Recapitulating equations above, we easily can see that:

> **If I = 1**, the field equation has an analogically classical relativistic expression:
>
> $$G^{\alpha\beta} = kT^{\alpha\beta}$$

So, we can consider that:

> **The relativistic classical field equations exactly describe only the special case of the space-time curvature in the presence of isolated gravitational field of the massive bodies.**

The relativistic field equations represent a reasonable approximation of DGIT, for the special case of the laminar isotropic (non-inductive) field, as well as Newtonian equations represent a reasonable approximation of the relativism for the special case of the light speed.

The accuracy of the equations could be improved by the scientists who exactly understand the relativistic space-time field; but I think if the mathematical formalism rejects the plain explanations and excessively mystifies and abstracts the real pure phenomena, we risk to imagine a false, misunderstood and illusive Universe.

Final Considerations

According to DGIT, gravitational field is an inductive gravitodynamic one, which seems to have a similar behaviour to a long-distance residual EM field.

DGI influences the massive bodies' motion and their position in the gravitational field, as well as the electromagnetic induction influences the dynamics of electric charges in an EM field.

DGI briefly can be described having the characteristics below:

1. DGI does not confute classical Gravity in the conservative balance.

2. DGI has huge values and shrinks the space inside the galaxies.

3. DGI tends to infinite in the center of the galaxies, forming black holes.

4. DGI generates the conservative balance at the edge of the galaxies.

5. DGI severely decreases on large distances, allowing the intergalactic space dilation.

DGI adjusts the general accepted Universe's forces, eliminating the necessity of some huge quantities of alleged Dark Matter in the pure architecture of the Universe.

Or, if it exists, it is not at all non-interactive with so called "Visible Matter", but just an unobserved part of this.

DGI adjusts Gravity, together creating a fine perpetual balance.

In the absence of DGI, the galaxies would disintegrate or would be gradually swallowed by their central black holes.

DGI is that Fibonacci's string, which arranges the matter in huge spirals that successively unfold on the vast spaces of the Universe.

The adherence of the Space to itself is a consequence of DGI.

Fine! The Space could be Tesla's Ether!

> **Dynamic Gravity Induction**
> **is**
> **The Missing Part.**

ANNEXES

ANNEX 1
Conservative and Inductive Gravitational Balance

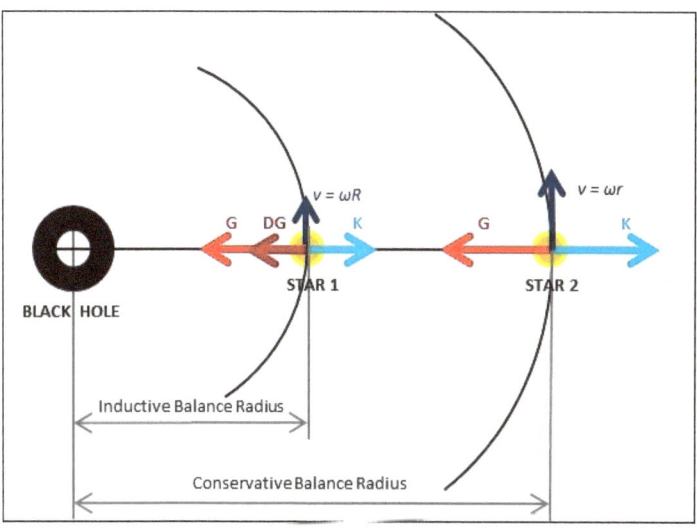

ANNEX 2
Graphic Representation of DGI and related characteristics in the Universe

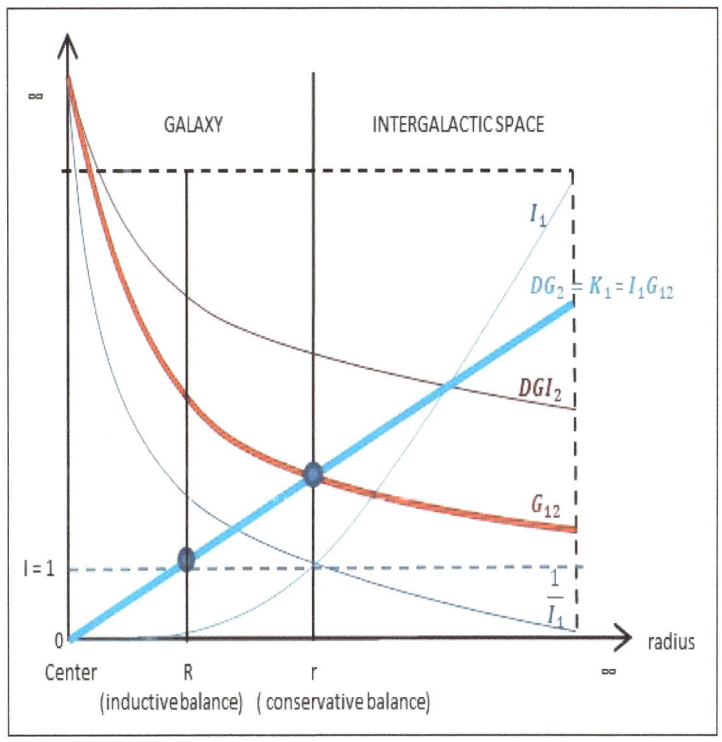

ANNEX 3
Graphic Representation of DGI and related characteristics inside of a Galaxy

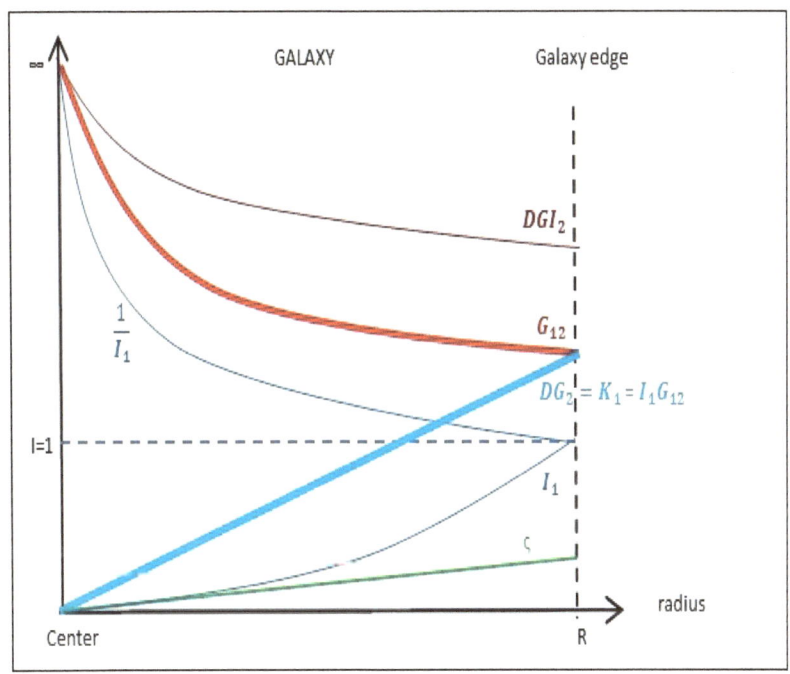

ANNEX 4
Graphic Representation of TOTAL DGI and related characteristics inside of a Galaxy

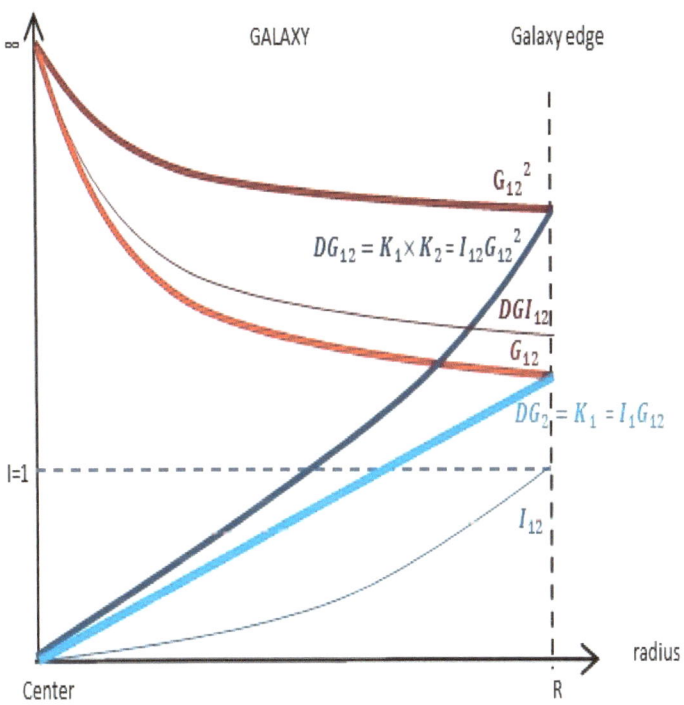

ANNEX 5
Graphic Representation of Real Mass and Apparent (Inductive) Mass in the Galaxy

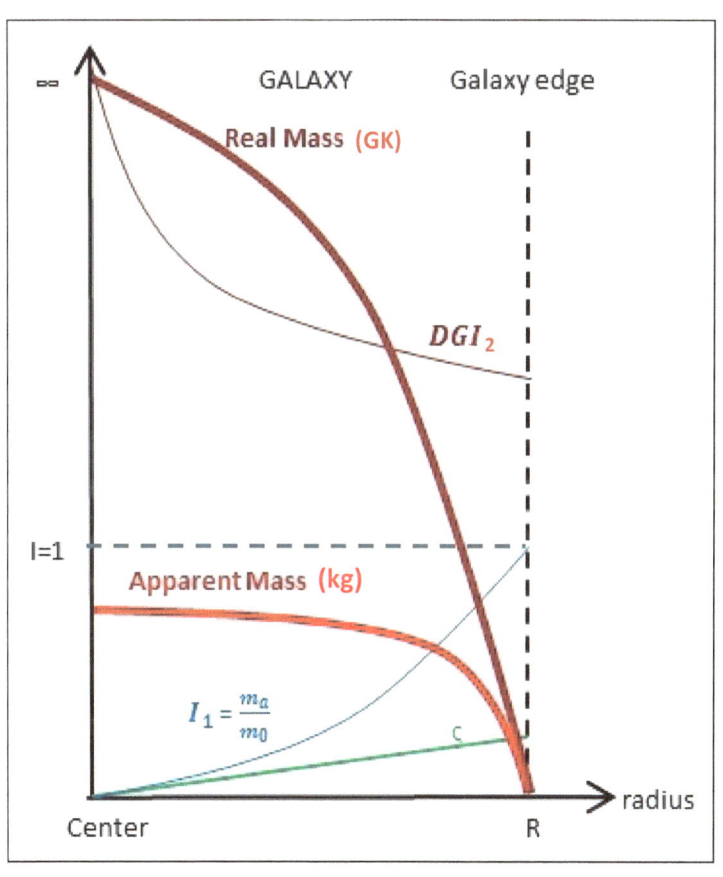

ANNEX 6
Graphic Representation of DGI and related value in a compound inductive gravitational field

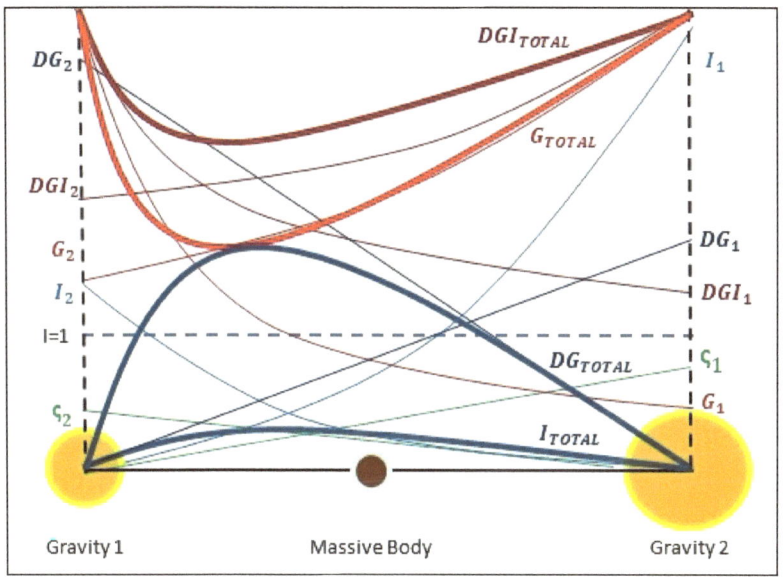

ANNEX 7
The spiraled shape of the Galaxy in accordance to Local Inductive Elasticity Ratio

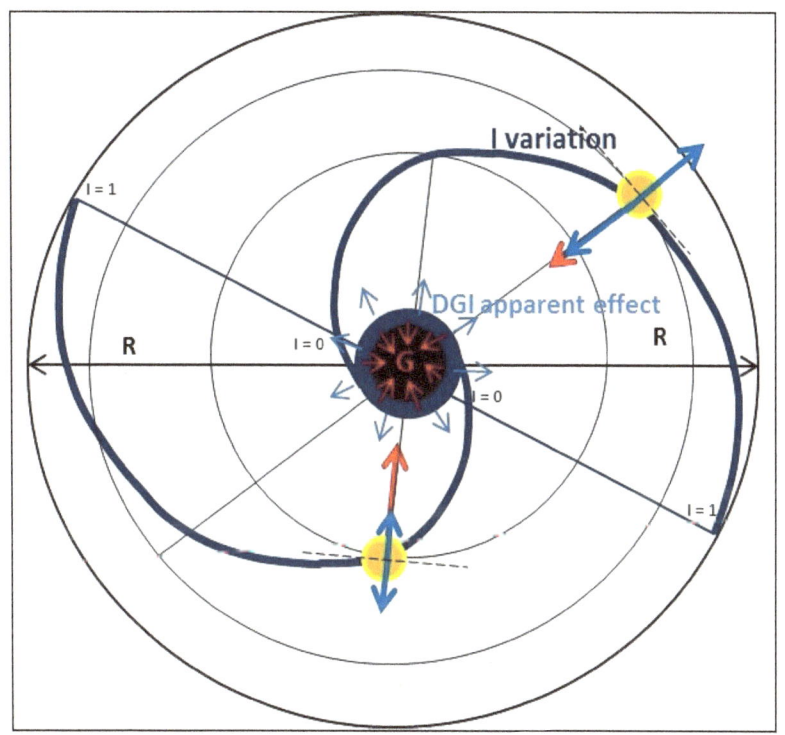

ANNEX 8
Universal Energy Systems Models

REFERENCES

DGIT represents a section of STEMIONICS, which is a personal Unified Universal Force Model (a Theory of Everything), yet in a living evolution.

According to STEMIONICS, the Universes are continuously generated, the origin of the Multiverse Matrix being a White Hole (a Noullon), constitued by STEMIONs (Space Time Enery Mass Information Originar Nodes).

Our Universe is an Energy Megaquanta, moving between the previous and the succesor Universe; all the Universes and the events occuring inside them, seem to be simultaneous for an observer which is situated in the Central Nullon.

The Space-Time represents the support field for all the electromagnetic interractions, and the architecture of this field is described in the DGIT-relativistic model.

Being captive inside a default space-time field, in order to preserve their Energy - Momentum, the Energy Quanta have to establish complexe EM interractions, partially described in

the Quantic model.

These interactions themselves, generate in their turn, the residual inductive gravitodynamic field, which is responsable for the Space-Time architecture.

However, I assume that DGIT could improve STEMIONICS model, in order to further develop an accurate model for the behaviour of small particles and EM waves.

Previous presentations of STEMIONICS Model:

1.   Cupşa O.S.,*"STEMIONICS – Grand Unified Theory"*, Ed. Nautica, Constanţa, 2012, 44p.
ISBN: 678-606-8105-74-1

2.   *"STEMIONICS – Grand Unified Theory"*, - poster and open session, Universitas Carolina Prague, Conference *"Relativity and Gravitation, 100 Years After Einstein In Prague"*, June, 25 – 29, 2012.

3.   Abstract *"STEMIONICS – A Grand Unified Theory Pattern"* in *"Book of Abstracts"* of the Conference *"Relativity and Gravitation, 100 Years After Einstein In Prague"*, p.78.

4.   Cupşa O.S., *"STEMIONICA – Marea Teorie Unificata"* (romanian language), Ed. Nautica, Constanţa, 2013, 740p.
ISBN: 978-606-681-003-6

INDEX

PERSONAL ANNOTATIONS

The Missing Part

www.ingramcontent.com/pod-product-compliance
Lightning Source LLC
Chambersburg PA
CBHW040904180526
45159CB00010BA/2918